Tools for Success:
Critical Skills for the
Construction Industry

Trainee Workbook
Third Edition

Prentice Hall
Boston   Columbus   Indianapolis   New York   San Francisco   Upper Saddle River
Amsterdam   Cape Town   Dubai   London   Madrid   Milan   Munich   Paris   Montreal   Toronto
Delhi   Mexico City   Sao Paulo   Sydney   Hong Kong   Seoul   Singapore   Taipei   Tokyo

National Center for Construction Education and Research

President: Don Whyte
Director of Curriculum Revision and Development: Daniele Stacey
Tools for Success Project Manager: Patty Bird
Production Manager: Tim Davis
Quality Assurance Coordinator: Debie Ness
Editors: Rob Richardson and Matt Tischler
Desktop Publishing Coordinator: James McKay
Production Assistant: Laura Wright

NCCER would like to acknowledge the contract service provider for this curriculum:
EEI Communications, Alexandria, Virginia.

This information is general in nature and intended for training purposes only. Actual performance of activities described in this manual requires compliance with all applicable operating, service, maintenance, and safety procedures under the direction of qualified personnel. References in this manual to patented or proprietary devices do not constitute a recommendation of their use.

10 9 8 7 6 5 4 3 2 1
ISBN 0-13-610649-8
ISBN-13   978-0-13-610649-4

Contren® Learning Series

Preface

To the Trainee:

Welcome to the world of construction! Construction is one of the nation's largest industries, offering excellent opportunities for high earnings, career advancement, and business ownership. Construction workers are also highly skilled and are usually specialized in one field, such as carpentry, electrical, or plumbing.

This book will help you on your way to a successful and fulfilling career in the construction industry by focusing on critical skills. After completing this course, you will be able to fill out a job application, write a resume, communicate effectively, build a strong relationship with your boss and co-workers, and recognize and deal with a number of workplace issues. These skills will be useful as you move along in your construction career. They will help you get the job that you want, and will make you a valuable employee.

New with this Edition of *Tools for Success*

NCCER and Pearson are please to present the third edition of *Tools for Success: Critical Skills for the Construction Industry*. One of the major changes to this book is the addition of "First Impressions: Getting a Job". This module will help your trainees learn how to fill out an application, write a resume, and prepare for the interview process.

Many other modules have been updated. We are pleased to be adding a note on generational diversity to the "Diversity in the Workplace" module. Communication skills have been updated in both the "Communication Skills I: Listening and Speaking" and "Communication Skills II: Reading and Written Communication" modules. Personal budgeting information has been added to help with "Managing Stress on the Job".

We invite you to visit the NCCER website at www.nccer.org for the latest releases, training information, newsletter, and much more. You can also reference the Contren product catalog online at www.nccer.org. Your feedback is welcome. You may email your comments to curriculum@nccer.org or send general comments and inquiries to info@nccer.org.

Contren® Learning Series

The National Center for Construction Education and Research (NCCER) is a not-for-profit 501(c)(3) education foundation established in 1995 by the world's largest and most progressive construction companies and national construction associations. It was founded to address the severe workforce shortage facing the industry and to develop a standardized training process and curricula. Today, NCCER is supported by hundreds of leading construction and maintenance companies, manufacturers, and national associations. The Contren® Learning Series was developed by NCCER in partnership with Pearson Education, Inc., the world's largest educational publisher.

Some features of NCCER's Contren® Learning Series are as follows:

- An industry-proven record of success
- Curricula developed by the industry for the industry
- National standardization, providing portability of learned job skills and educational credits
- Compliance with the Office of Apprenticeship requirements for related classroom training (*CFR 29:29*)
- Well-illustrated, up-to-date, and practical information

NCCER also maintains a National Registry that provides transcripts, certificates, and wallet cards to individuals who have successfully completed modules of NCCER's Contren® Learning Series. *Training programs must be delivered by an NCCER Accredited Training Sponsor in order to receive these credentials.*

[Note: *Tools for Success* is not a standardized craft; therefore, the NCCER National Registry does not offer credentials for completion of this title.]

Contren® Curricula

NCCER's training programs comprise over 50 construction, maintenance, and pipeline areas and include skills assessments, safety training, and management education.

Construction and Maintenance

Boilermaking
Cabinetmaking
Carpentry
Concrete Finishing
Construction Craft Laborer
Construction Technology
Core Curriculum: Introductory Craft Skills
Drywall
Electrical
Electronic Systems Technician
Heating, Ventilating, and Air Conditioning
Heavy Equipment Operations
Highway/Heavy Construction
Hydroblasting
Industrial Coating and Lining Application Specialist
Industrial Maintenance Electrical and
    Instrumentation Technician
Industrial Maintenance Mechanic
Instrumentation
Insulating
Ironworking
Masonry
Millwright
Mobile Crane Operations
Painting
Painting, Industrial
Pipefitting
Pipelayer
Plumbing
Reinforcing Ironwork
Rigging
Scaffolding
Sheet Metal
Site Layout
Sprinkler Fitting
Welding

Pipeline

Control Center Operations, Liquid
Corrosion Control
Electrical and Instrumentation
Field Operations, Liquid
Field Operations, Gas
Maintenance
Mechanical

Safety

Field Safety
Safety Orientation
Safety Technology

Management

Introductory Skills for the Crew Leader
Project Management
Project Supervision

Supplemental Titles

Applied Construction Math
Careers in Construction
Tools for Success
Your Role in the Green Environment

Spanish Translations

Acabado de Concreto, Nivel Uno
Aislamiento, Nivel Uno
Albañilería, Nivel Uno
Andamios
Carpintería
    Introducción a la Carpintería, Nivel Uno
    Carpintería de Formas, Nivel Tres
Currículo Básico:
    Habilidades Introductorias del Oficio
Electricidad, Nivel Uno
Herreria de Refuerzo, Nivel Uno
Instalación de Rociadores, Nivel Uno
Instalación de Tubería Industrial, Nivel Uno
Orientación de Seguridad
Principios Básicos de Maniobras, Nivel Uno
Seguridad de Campo

Acknowledgments

This curriculum was revised as a result of the farsightedness
and leadership of the following sponsors:

ABC of Southern California
Alfred State College
Becon Construction Co., Inc./Bechtel
H.B. Zachry Co.
IMTI of Connecticut
IMTI of New York
Kennco Plumbing
River Valley Technical Center

This curriculum would not exist were it not for the dedication and unselfish energy of those
volunteers who served on the Authoring Team. A sincere thanks is extended to the following:

Alberta Casey
Erin M. Hunter
Melissa Malone
Janet Marhefka
Thomas G. Murphy
Glen O'Mary
Ray G. Thornton
Marcel Veronneau

NCCER Partnering Associations

American Fire Sprinkler Association
Associated Builders and Contractors, Inc.
Associated General Contractors of America
Association for Career and Technical Education
Association for Skilled and Technical Sciences
Carolinas AGC, Inc.
Carolinas Electrical Contractors Association
Center for the Improvement of Construction
    Management and Processes
Construction Industry Institute
Construction Users Roundtable
Design Build Institute of America
Green Advantage
Merit Contractors Association of Canada
Merit Shop Training, Inc.
Metal Building Manufacturers Association
NACE International
National Association of Manufacturers

National Association of Minority Contractors
National Association of Women in Construction
National Insulation Association
National Ready Mixed Concrete Association
National Systems Contractors Association
National Technical Honor Society
National Utility Contractors Association
NAWIC Education Foundation
North American Crane Bureau
North American Technician Excellence
Painting & Decorating Contractors of America
Portland Cement Association
SkillsUSA
Steel Erectors Association of America
U.S. Army Corps of Engineers
University of Florida
Women Construction Owners & Executives, USA

Contents

First Impressions: Getting a Job

"People might not get all they work for in this world, but they must certainly work for all they get."

– Frederick Douglass,
American statesman

Introduction

You want to find a job in construction. Good! While you should be confident that you'll find one, you should know that nothing's guaranteed. Getting a job in construction has its challenges, but with proper preparation, you should be successful.

This module discusses the four major components of a job search:

1. Searching for construction jobs effectively and efficiently.

2. Filling out a job application properly.

3. Putting together a useful resume of your skills and experiences.

4. Preparing for and then participating in a job interview so that you make a positive first impression on a potential employer.

You'll need to be familiar with each step, so that you can uncover available and suitable positions in construction, let an employer know you want a particular job, and impress that employer enough so that you'll be hired. If you know how to get the most out of these four steps, your job searches will be more rewarding while looking for and finding a job.

Your instructor is a reliable resource for further help and can recommend media tools to use.

Job search

You have a number of options to help you search for construction jobs. There's the old standby of going through the help-wanted ads in the newspaper. That's still an effective way to look for construction work, but there are other successful and proven methods you should use, including the internet, word-of-mouth, and temporary-help agencies.

The internet has made looking for a job easier than before. (If you don't own or have access to a computer at home, you can go to the local public library, where you can sign up to use a computer and get on the internet. It's usually free.) You can use the internet both to find available jobs and to market yourself to let employers know who you are and why you're a good potential employee. To get the most out of using the internet in your job search, follow these guidelines:

- Employment websites often have online databases of available jobs. These websites sometimes offer other services for job seekers as well. The NCCER Careers website, *careers.nccer.org*, is one excellent resource that covers the construction industry. You can find job-search boards at large general-purpose websites such as CareerBuilder, as well. Both types of websites can be useful, but the job-search boards on the niche websites are probably easier and quicker for you to use when you are looking for construction work.

- Online bulletin boards about the construction industry are another good source of jobs, especially those available in your specific area.

- To access job listings from both bulletin boards and more general-purpose employment websites, you'll likely have to register with the website's sponsor. Registration is usually free, but sometimes you'll have to complete an online application, especially if you want to post your resume online.

- When looking over a job-search website, look for the words *construction laborer* or *general construction worker*.

- The internet is a great way to network. You can do this in a number of ways. Start a blog about construction. Use a social networking account to promote your skills and experience, and include a picture of yourself. Be careful what you put on your site, though; anyone can access it.

Word-of-mouth is another great way to uncover job opportunities. Some people think the best construction jobs are found through word-of-mouth. How does this method work? Tell everyone you know—not only family and friends, but neighbors, and even the person who delivers your mail—that you are looking for construction work, and then ask them to tell everyone *they* know. You might be surprised at the response. If you know someone who has contacts in the construction industry and that person gives you a good recommendation, that's a good reference.

Temporary agencies generally have a roster of construction workers whom the agency sends to companies that need short-, medium-, or long-term help. The agency might specialize in construction employment, or it might serve a number of different industries. On its roster, the temporary agency might have representatives of many different construction trades, such as bricklayers, heavy machinery operators, or concrete finishers, or the agency may specialize in one or two particular trades. You should look for agencies that specialize in providing entry-level workers. If your temporary-employment agency has assigned you to a construction company, your boss will likely be an employee of that company and not of your agency. Even though you may technically be an employee of the temp agency, the supervisor is still your boss. That means you must listen closely when your supervisor is speaking and do what you are told.

One reason you should use a number of different job-search resources and strategies is that construction work is cyclical, which means the availability of construction jobs can depend on economic or weather conditions. There will be times when jobs are readily available and times when they aren't. During those slower times, the more job-search resources you know how to use, the better your chances of finding a job. When the

> "I'm a great believer in luck and I find the harder I work, the more I have of it."
>
> – Thomas Jefferson

economy is sluggish, there tend to be fewer construction projects and thus fewer jobs available. Because so much construction work is outdoors, weather is a big factor, which means there may be fewer jobs during the winter. The important thing is not to get discouraged when things are slow and to recognize that finding a construction position may be influenced by factors beyond your control. If things are a bit slow now, they will get better; the economy will improve, the weather will get warmer. Don't allow yourself to get discouraged. Perseverance pays off!

Two other things to keep in mind if your job search isn't quite as productive as you want it to be:

- Take a good look at the companies to which you're applying. Are these companies that generally hire someone like you, who is just starting out? Or do they primarily employ highly skilled craftworkers like plumbers and electricians? If that's the case, it might not be worth your time or effort to apply to those companies.

- Look for work out of state, if you can relocate temporarily or permanently. More jobs might be available in other states.

Job application

Whether you fill out a job application online or on paper, there are guidelines to follow:

- Read the application carefully before you start filling it out, and follow all instructions.

- Locate your former employment data, which might be found on such items as old check stubs, past-year tax records, or previous resumes. If you're filling out an application prior to a job interview, take this information with you to the interview so that it's readily available when you fill out your job application.

- If you're filling out a paper application, print legibly and think about what you want to write before you put it down. You don't want your form to be messy or hard to read.

- Avoid using abbreviations when filling out the form; the reviewer of your application may not know what they mean.

- Fill out the application accurately and completely. Do not leave an item blank; if the item does not apply to you, write N/A (meaning "not applicable") in the space. Make sure all company and supervisor names are correctly spelled, and that phone numbers and addresses are accurate and up-to-date.

- Answer all questions with complete honesty. If you have a criminal offense, be upfront about it. You can be fired if your employer finds out you lied on your job application, even after you've been at the job for awhile. Lying includes deliberately omitting certain details as well as making false statements. Remember, everything you write on the form is subject to verification by the company, and if it discovers anything false or deliberately misleading on your application, even about something you think is minor, your application will likely be rejected.

- Do not exaggerate when filling out your application. If you list responsibilities you didn't have and the company subsequently discovers this, you can be terminated from your employment.

- Try not to leave any blanks when filling out your employment history. Be aware that during an interview, you may be asked about gaps in your employment record.

- If you have no prior work experience, make sure your education information is complete. Don't try to manufacture work experience. Think of other experiences in your life when you were entrusted with responsibilities or were held accountable for what you did, and put those experiences down.

- Make sure to list anything you're currently doing to further your education or abilities, such as classes you're taking or certifications you're pursuing. Be sure to list any certifications (such as NCCER's) you have.

- If the application has a space for salary expectations or salary requirements, answer "open" or "negotiable."

- At the end of your job application, you'll be asked to confirm that all the information you provided is, to the best of your knowledge, correct and accurate. Your signature will be your confirmation. Before you sign, review all the information on the application for accuracy and completeness.

Filling Out a Job Application When You Have Limited Experience

Many people, especially when they are just starting out, feel at a disadvantage because they don't have much, if any, actual work experience or work-related references. Don't worry; there are ways around this. If you are just beginning your work career and lack real work experience (that is, full-time paid work), here are some things you should include on your application:

- **Summer jobs.** If you worked for anyone during your summer vacation(s), be sure to list what you did and who you worked for on your application. Summer jobs, just like full-time jobs, require you to show up for work on time and do your job responsibly and well. Don't overlook work you may have done for a relative or a neighbor.

- **Skills.** Be sure to list every tool and machine you know how to operate, including such devices as photocopiers, calculators, cash registers, or computers. Think about any tasks you've done and every piece of equipment you've used safely and well during those tasks, and list those devices on your application. Don't neglect to mention other abilities. For instance, if you think you're a good communicator or that you're well organized, list those skills as well—if you can cite specifics. For instance, how did you demonstrate your communication skills when you worked with the public at your summer job? What about your volunteer work proves that you're well organized?

- **Volunteer work.** This type of work gives you experience you can use on a future job, and it demonstrates that you are a responsible person. Examples of volunteer projects include Habitat for Humanity, food or clothing drives, and youth group projects.

- **References.** References are particularly important if you have little or no work experience. References can be from anyone you worked for over summer vacation, and they can also be from a volunteer coordinator, a school or youth group counselor, your teachers or instructors, a coach, or a religious advisor. You should not list relatives as references. You should have at least three to five references. Make sure you get the individuals' permission to use them as references, and be certain they will speak highly of you if contacted by a potential employer.

Filling Out Online Job Applications

When filling out an application over the internet, you should follow all the tips for filling out a paper application form. There are also some special considerations you'll need to be aware of when filling out an application online:

- If you are interested in applying for work at a particular company, see if that company has a website. If it does, look for a link that says Job Application; usually, these links can be found on the web page that covers job opportunities. Check to see whether you should apply for a specific position (for instance, Construction Laborer); if there is such a link, clicking on it will usually take you to the company's online application for that position.

- Fill in all the necessary personal information: your name, date of birth, phone number, street address, e-mail address, and Social Security number. Many online applications cannot be submitted unless all this information is provided.

- If the online application has a section where you can upload your resume (which means it'll be sent along with your application to the company), it's a good idea to provide it, even if your resume contains basically the same data you provide on the online application.

- Double-check your application before submitting it. Check spellings of names and companies, phone numbers, and addresses. Remember, once an online application is sent out to a company, it cannot be recalled for editing. Be particularly careful if the application includes multiple-choice questions; make sure you've marked the answers you want before you send the application off.

- Be sure to save your data if you're allowed to by the computer software used with the online application. Try to complete the application during one sitting, especially if you cannot save what you've entered as you go along. Before you start filling out the online application, consider whether you have the time to finish it at one sitting; if you don't think you do, fill it out at another time.

- Most companies will send you an e-mail to confirm that they've received your application. This will let you know your application is being processed. Save this e-mail for future reference.

- You can find examples of online application forms on the internet; consult your instructor or type in "sample job application" in an internet search engine.

Resume

One of the best ways to give an employer useful and relevant information about you is through a resume. In the past, those looking for entry-level construction jobs didn't need a resume. Today, a well-organized and informative resume gives you a competitive edge. This section discusses how to create such a resume.

Check that your contact information is correct. This may seem obvious, but a common problem with resumes is incorrect or missing personal information. Make sure you check the accuracy of all your essential personal data: address, phone number, and email address or cell phone number if you have them.

Display your personal information prominently. Put your full name at the top of your resume and don't use nicknames. Your name should appear slightly larger than the rest of the resume's content. Underneath your name, list your street address and phone number.

Organize your resume so employers can easily find what they're looking for. There are a number of different theories on how to do this best. One logical way is by arranging your resume's content by these sections:

- **Objective.** This should be the first section of your resume, located immediately below your name and contact information. Here you state exactly what kind of job you're seeking. This should be a concise statement, only three or four lines long. If you're having trouble coming up with something short and pertinent, borrow some of the language used in the job posting to which you responded. You can also mention your career goals here, as long as you keep the statement brief.

- **Skills and Qualifications.** This should be the next section of your resume. Here you list exactly what you know how to do on a construction site, including all the construction tools and/or equipment you know how to use. You may want to include a line about when and how you learned to use that equipment. You should use bullet points (like those used in this list) to present this information. You should list your skills and qualifications before you present your job experiences, because at this stage of your career, potential employers will be more interested in what you can do and not where you have worked.

- **Job Experience.** This section comes after Skills and Qualifications. In this section, list your past employment. Put your most recent job down first. For each job, indicate the title of that position, your employer, and how long you worked there. With each job, you may add a line on what you accomplished there and what skills you used. Put down *any* construction-related experience you might have; it doesn't necessarily have to be from a formal job. If you helped a relative to build a deck or a neighbor to refurbish a basement, put that in your resume; for someone at your stage, it counts as construction-related experience as much as if you'd done it while employed by the Bigwig Construction Company.

- **Education.** In this next section, you'll let prospective employers know when and where you went to school, or if you're still attending school. Include the names of the schools you attended, when you went there, whether or not you graduated, and the type of degree or diploma you obtained. If you've taken (or are taking) any certification courses, or if you've received (or are receiving) any special construction-related training, don't forget to put those down.

- **Other Skills.** You don't have to include this section, but it's a good place to mention any other skills you may have, such as the ability to work well in a team or experience in working with the public. Keep in mind that if you mention these skills in your resume, you're liable to be asked about them during a job interview. Make sure you can provide specific examples of how you've applied any skills that you list.

Use a proper format for your resume. A common complaint among companies is that the resumes they receive have elaborate or confusing formats. Make certain you present your information in a clear, concise fashion. Organizing your information using the section titles just discussed will help.

Keep your language simple. Don't use fancy words in an attempt to make your skills and experiences appear more important. Avoid unnecessary adjectives. Be direct in how you phrase things.

Be honest. Never provide misleading or false information in a resume. Don't lie about your degree, your skills, or your experience. Don't exaggerate former accomplishments or change dates in an attempt to cover up periods when you weren't employed or in school. Don't claim that you can do things you can't.

Round out your resume. If you don't want to list your references at the end of your resume, make sure you state instead, "References available upon request." Make sure you've prepared those references. If you're in a position where you can move to another part of the country, express your willingness to relocate temporarily; the availability of construction work can vary from region to region and season to season.

Keep your resume short. Try to limit your resume to one page. It should never be more than two pages. If it goes to two pages, use separate pieces of paper; use one side of a piece of paper only.

Check and recheck your resume before you send it out. Make sure there are no spelling or grammar mistakes. Get someone else to look at your resume to make sure that it is easy to understand.

Sample Resume Format

Your Name

Address

Phone Number

Email Address

Employment Objective

Seeking a construction job where I can use my carpentry and custom woodworking skills.

Experience

From May 2000 to present. Journeyman carpenter. LJL Construction, 123 Hammer Heights Rd., Fairfax, Va. 800-222-2222. Built custom cabinets, closet systems; framed single-family homes.

From June 1996 to May 2000. Carpenter. Hammer Company, 456 Lathe Lane, Philadelphia, Pa. 900-456-4566. Built closets and storage systems for commercial storage units.

From February 1994 to June 1996. Carpenter. Blue Ridge Builders, 789 Mountain Way, Fresno, Ca. 888-123-4567. Framed condo units, built fences and raised wood walkways.

Skills

Practical knowledge of carpentry hand and power tools, measuring tools, and woodworking tools. Qualified to read residential and commercial blueprints. Experience with both stick and modular frames, including steel frames. Working knowledge of roofing and residential plumbing. Trained and certified to use powder-actuated tools.

Construction Education

Apprenticeship. Carpentry. MK Builders Inc., Fresno, Ca. Certificate, 1996.

Professional carpentry certification program. ABC Training Corp., Philadelphia, Pa. Certificate, 2000.

Powder-actuated tools certification program. All States Community College, Fairfax, Va. Certificate, 2001.

Job interview

The job interview is your opportunity to sell yourself. It's a crucial step to being hired by a construction company or by a temporary agency that sends workers out to construction companies. The interview is when you make your impression as a potential worker and as a person. Decisions on hiring are often heavily based on the interview, on the answers you give and the overall impression you make.

A poor performance on an interview will sink your chances at working for that company. It's important, therefore, to plan and prepare for your interview. The better prepared you are, the more confident you'll act during the interview, the clearer your answers will be, and the better an impression you will make.

Before the interview

A successful job interview starts with preparation: practicing what you'll say, knowing exactly when and where the interview is scheduled, getting everything ready the night before, and arriving on time and looking presentable.

Practice, practice, practice. The more you practice before the interview, the more comfortable you will feel while it is taking place.

- Rehearse what you intend to say during the interview, both your responses to questions and the questions you'll ask.

- Practice with a friend, a family member, or in front of a mirror.

- Find out what the company does and what you'll be doing if you're hired.

- Practice doesn't mean over-rehearsing. You shouldn't try to memorize all your answers to every conceivable question. (One thing you need to memorize, however, is the name and title of the person who'll be interviewing you; always refer to this person by using Mr. or Ms. unless they tell you otherwise.)

Location, location, location. You do *not* want to be late for a job interview. Being late makes a poor first impression, and tells the company you're not serious about getting the job.

- Know exactly where the interview is to be held and how long it will take you to get there.

- Take a practice run—whether you are planning on driving to the location or taking public transportation—at the time of day your interview is scheduled.

- Allow enough time for traffic or bus delays. If you find it takes 30 minutes to drive there on your practice run, you might want to allow an hour on the day of the interview. If it takes an hour to get there by public transportation on your dry run, you may want to leave two hours before the interview starts (especially if you have to switch buses or trains).

- If you're driving, find out about the location and availability of parking; call the company to find out. The company may have onsite parking for visitors, which you can use.

Get everything you'll need at the interview ready the night before. The idea is to get as much done as possible before the big day; the less you have to do before the interview, the less stressed-out you'll be when the interview starts.

- Gather everything you'll need to take to the interview: your identification, references, resume, certifications, any notes to yourself. That way, you'll avoid having to search for important papers at the last minute or right before you leave the house. Refer to the upcoming section on what to bring on an interview for the specific items you'll need.

- Make sure your job references are in order.

Get there early, and look respectable. First impressions matter, and that means not being late or showing up looking unkempt.

- Dress properly. Make sure your clothes are clean, neat, and not ripped. The rule with job interviews is to dress for the job you want.

- Pay attention to all aspects of your personal appearance and hygiene; shower, wash your hair, and brush your teeth before you leave for the interview. Make sure you are well-groomed; hair should be neatly cut and beards should be trimmed.

- Arrive at least 15 minutes before the interview is scheduled to start. That will give you time to use the bathroom, check your appearance in the mirror, and collect your thoughts.

- If you are welcomed by a receptionist or assistant, greet them with courtesy and respect. That's where you make your real first impression.

- If you are given a job application, fill it out neatly, completely, and accurately.

- Relax. If you're prepared and know what to do and say and how not to act, your interview will go well and you'll make a positive impression.

What to bring on an interview

You'll need to bring a number of things to your interview. You'll probably fill out an application form before the session begins, so you'll need to have everything required to fill it out completely, including the following items:

- At least two forms of official identification. Bring your driver's license and your Social Security card. A Social Security number is a necessity. If you don't have a driver's license, you'll need to provide your birth certificate or passport (if you have one). If you don't have a Social Security card, you'll need to bring your visa or a passport (something with your Social Security number on it).

- A list of all of your essential information (such as previous jobs, schools, or any other skills or experiences); you may forget some of these details as you're filling out the job application.

- Your references.

- Copies of your resume, in case the interviewer asks for it.

- A notepad. Before you start using it during the interview, however, ask the interviewer's permission to write things down. The interviewer may not allow it, but the fact you asked to use it won't be held against you.

- It's probably acceptable to bring along notes you've written down to help you during the interview; for example, a list of questions you want to ask about the job and the company. Still, you should check with the person who'll be conducting the interview before the session starts to make sure it's all right. Do not bring a lot of extensive notes to the interview, however. The interviewer wants to have a discussion with you, not listen as you read a prepared statement.

- If you have any questions before the interview, call or e-mail the contact person who arranged your interview.

Certifications

A certification is a confirmation that the person who possesses it has the skills or knowledge to complete a certain job. There are a number of certifications related to the construction industry, and some of these are increasingly becoming requirements for employment. One such certification comes from completing the Occupational Safety and Health Administration's (OSHA) 10-Hour Construction Outreach Training Course. Other certifications are in cardiopulmonary resuscitation (CPR) and basic first aid. NCCER offers a certification in introductory craft skills, which covers, among other topics, safety, communication and employability skills, and the basics of blueprints, construction math, and tools. Before your interview, call the company to find out if the position you are applying for requires any certifications.

During the interview

Each job interview you go on will be different; interviewers look for different things and will ask different questions. During any interview, however, the following good practices will help you to make your best impression.

Shake hands with the interviewer properly. Don't offer a limp handshake. Your handshake should be firm, but you shouldn't be trying to crush the interviewer's hand, either.

Wait until you're offered a chair before you sit down. This is the polite thing to do, and it gets things off to a good start.

Stay aware of your body language during the interview. This doesn't mean thinking about your body language so much that you lose your focus. It does mean to look alert and sit up straight at all times, and to make eye contact when you answer the interviewer's questions.

Use an appropriate tone of voice. A forceful voice projects confidence; speaking softly doesn't. If you speak too quietly or not clearly, the interviewer might think you won't be heard at a hectic, noisy jobsite. This doesn't mean you should shout or yell, though.

Use concise, clear language. Try to limit your use of *um* and *uh*. Make your points using the facts you've prepared. Avoid using slang as much as you can. If a question temporarily throws you off, don't babble or stammer if you don't have an answer ready. A helpful tip is to ask the interviewer to go into a little more detail about the question, which will give you a moment or so to mentally prepare an answer. It's fine to take a second or two before answering a question.

Take the proper tone. Talk about what you'll bring to the company and not about what you want the company to do for you. Talk confidently about your abilities and your strengths; this is not the time to be humble. Show enthusiasm when talking about the job. However, this doesn't mean you should boast about yourself.

Ask questions. Asking questions shows that you're interested in the job. Do some research about the company and the position before the interview, so that the questions you ask have some substance. You don't want to waste the interviewer's time by asking such things as, "Do we have to work outside when it's cold?" The more substantial your questions are, the more it will show the interviewer that you're serious about the interview and the position. It helps if you prepare a few questions before the interview, such as:

- What will the company expect of me if I'm hired? What are the specific tasks I'll be told to do?

- What is the company culture like? Are people informal or formal with each other on the job? Do the workers hang around a lot with each other after work, or do they generally go their separate ways?

- Does this job or this company offer apprenticeship opportunities, or other types of on-the-job training, which will allow me to learn the skills I'll need in my construction career?

- What opportunities will I have to take on more responsibilities?

Stick to the truth, the whole truth, and nothing but the truth. This means more than not lying. Being honest during an interview means not making up stories or stretching the truth about what you've done in previous jobs. It also means saying exactly what you did, even if you think it doesn't sound all that impressive. If your last job was sweeping up at the local chain restaurant, don't try to make it sound as if you were disposing of hazardous materials or in charge of cleanliness engineering for a worldwide firm. If you did a good job at it and can show you gave a honest day's work for your pay, that'll be a fine answer. At this stage of your career and for the kind of jobs for which you're applying, no one is expecting you to have the experience or abilities of a four-year journeyman.

Be prepared for questions. Although every interview can be a bit different, most involve some standard questions. If you've prepared well-organized responses to these questions, you'll not only make a favorable impression, but you'll gain confidence that will help you in answering other questions. See the next section on challenging questions often asked in interviews for more on the types of questions you can expect and the kinds of answers you might want to consider.

Conclude the interview strongly. Thank the interviewer for his or her time and interest. Restate your interest in the position. Ask when it would be appropriate for you to get back in touch with the company, if you don't hear from them first. Even if you think the interview didn't go well, make sure you remain positive and leave, as you entered, with a smile on your face.

Follow up. If the company hasn't contacted you by the time they said they would, call the company.

Don't get discouraged. If you don't hear back or you don't get the job, remember it's just one company. There are others. Consider it valuable experience, and keep trying.

Challenging questions often asked in interviews

Here are some of the more difficult questions you could be asked during an interview, as well as some possible answers:

Question: "Tell me about yourself."

Possible answers: What your interviewer wants to hear in response is a quick summary of who you are and why you think you should be hired. Talk about what you've done and how your past work is good preparation for the position. Use specifics. For instance, "My ambition is to become (list the construction specialty you want to pursue)." Or, "My work experience has been centered on improving my _____ skills."

Question: "Have you ever had any disagreements with your bosses?"

Possible answers: First off, don't answer "no." If you do, the interviewer will likely keep asking you about the topic to find some evidence of conflict. So don't be afraid to answer "yes," but focus on how you responded positively to the disagreement and what you did to resolve the conflict. A useful answer might be, "Yes, I've had differences with my bosses, but not major ones. I always try to understand what my bosses want of me and what their point of view is, and I try to make myself clear to them as well." Or, "I've had disagreements with my bosses, but I always try to resolve them and learn from them."

Question: "Where do you expect to be in your construction career five years from now?"

Possible answers: "During the next five years, I'd like to become the best _____ I can be." Or, "I'd like to have further responsibilities, such as _____"

Question: "What is your greatest weakness?"

Possible answers: Don't select a strength and try to present it as a weakness; don't answer this question by saying something like, "My greatest weakness is that I work too hard." That misses the point of the question and usually doesn't sound very convincing either. You want to answer this question by talking about a true weakness that you're working to overcome. "I need to be better organized" could be one answer; "I try to do too much at once" could be another. When answering this question, don't be reluctant to talk about a real weakness as long as you discuss it in terms of what you're doing to overcome it.

Question: "How do you feel about working as part of a team?"

Possible answers: You probably think the best answer to this question is, "I feel great about working as part of a team." That is the answer most people give, and interviewers won't necessarily be impressed if you just leave it at that. To give this response substance in the eyes of the interviewer, you'll need to be able to cite specifics demonstrating how well you've worked as part of a team in the past. Bring up examples from prior jobs that display your ability to be a good teammate; if you don't have examples from work, use your experiences from such activities as participating in organized sports or working on a school project.

Question: "Why should this company hire you?"

Possible answers: There's an obvious answer for this question, too—"Because I'm the right person for the job"—and once again, just about everyone will answer this question the same way. To make this answer credible, you should cite specifics that demonstrate those qualities that make you the best person for the job. For example, "I'm the right person because I'm a hard worker; at my last job, for instance, I put in a lot of overtime voluntarily." Or, "I'm the person you want for this job because I'm very responsible; for instance, my last company held me accountable for making deliveries to a warehouse, and I always did that on time."

Other Common Interview Questions

- "Why do you want to work in construction, and in particular for this company?"

- "What skills do you think this job requires?"

- "How did you hear about this company?"

- "What do you like or dislike about your current (or your last) job? Why are you leaving it (why did you leave it)?"

- "Did you enjoy school? Did you get your work done on time, or did you wait until the last moment?"

- "Give me an example of a major problem you faced and how you responded to it."

- "What accomplishments are you most proud of in your life? What are some of the things you're least proud of in your life?"

- "Do you have a criminal history or any issues with the law? Would they affect your ability to do the job?"

- "Random drug and alcohol testing is mandatory for this company's employees. Do you object to such testing? If so, why?"

- "During the past year, did you show up for work or school regularly, or were you often absent? Were those absences unexplained? How many days did you miss? Did you show up late to school or work regularly? If so, why?"

- "In your last job, did you work much overtime? Did you volunteer to work overtime? Are you willing to work overtime here?"

- "Have you ever been fired from a job? If so, why?"

- "What questions do you have about this company, or about this position?"

Be aware that there are some things an interviewer cannot legally ask you, such as about your marital status, ethnic background, or any other issue not related to the position. Such inquiries are not permissible because your answers could be held against you in a discriminatory fashion during the hiring process. Questions about your religious beliefs or political views are not appropriate, either. It's rare you'll come across this problem during an interview, but if you do, it is up to you whether or not you answer such questions. Keep in mind that if you don't answer because you sense that your reply will end your chances of getting the job, the interviewer and company might find some other reason not to hire you anyway. If you are asked inappropriate questions about your race, ethnicity, sexual orientation, or religious beliefs, or if you are asked questions such as "Do you think you are too old for this job?" or if you're a woman, "Are you planning on having kids any time soon?" chances are that's a workplace you'd feel uncomfortable in even if you got the job. You'd probably be better off looking for a job somewhere else.

What not to say or do during an interview

Some things that you might say or do during an interview could dramatically reduce—or even ruin—your chances at getting the job. By avoiding these mistakes, you'll make a much more favorable impression. *Never* do any of the following during the interview:

- Chew gum or smoke, or ask the interviewer for permission to do either.
- Tell jokes, even bland ones; you're not interviewing for a job at the local comedy club.
- Answer cell phone calls during an interview; if you have a cell phone with you, turn it off.
- Comb your hair or put on makeup.
- Name drop. If you have relatives or friends who work for the company (or work in the construction industry), don't talk about them. The interviewer wants to hear about you.
- Convey a negative impression to the interviewer through your body language:
  - Fidgeting in your chair or fiddling with an object, which could indicate to the interviewer that you lack concentration.
  - Slouching in your chair; this may indicate that you're not taking the interview seriously.
  - Speaking too closely to the interviewer's face; this could be interpreted as aggression.
  - Failing to look the interviewer in the eye; you don't have to stare at the interviewer during the whole session, but if you consistently avoid eye contact, that could be interpreted to mean you're bored, impatient, disdainful, or even lying.
- Talk too quickly or mumble.
- Try to dominate the conversation by talking too much or too loudly.
- Interrupt the interviewer.
- Say too little, or speak too softly. If you're silent or can't be heard, the interviewer won't be able to find out much about you.
- Bring up potentially controversial topics. An interview is no place to talk politics.
- Act as if you'd take any job. You may really need a job, but try not to show it during the interview. Sounding desperate won't do much for the positive, confident impression you're trying to leave.
- Brag or act pushy. Don't boast about yourself, especially about things that have nothing to do with your job.
- Bad-mouth your previous job or employer, or your former boss or co-workers. You'll come off as a tattletale or a backstabber.
- Bring up personal issues or family problems.
- Use profanity.
- Say anything that's offensive.
- Stretch the truth or lie.

Summary

Getting a job in construction can be a challenge, but if you go about it properly and if you persevere, you will find one. There are four primary components to getting a job. First, search for a construction position using the internet, word-of-mouth, and temporary agencies, as well as newspaper ads. Next, fill out the job application carefully, completely, and honestly. Third, put together an informative and concise resume, including an objective, skills and qualifications, and your education and previous work experience. Finally, take the time to prepare for the job interview, so that you can successfully sell yourself to a potential employer.

On-the-Job Quiz

Here's a quick quiz that allows you to apply what you've learned in this module. Select the best possible answer, given what you've learned.

1. You're compiling a resume and sending it to three companies that you are interested in. What is your best course of action?

   a. Print your resume on expensive paper so it will look really good.

   b. Use lots of words so that your resume fills more than two pages, making you look more experienced.

   c. Include bulleted lists, boldface type, italic type, a table, and a photo of yourself.

   d. Have a family member check it for mistakes before you make copies and send it out.

2. Your interviewer observes that you mentioned on your resume how well you work in a team. The interviewer asks you to elaborate. How should you respond?

   a. "I get along well with my brothers, and your family is kind of like a team, isn't it?"

   b. "I got a temporary job hanging drywall, working with a bunch of people I didn't know, and I got along well with them. We got the job done in time and under budget. Would you like the name of the person who supervised that work, to find out more?"

   c. "I played football in high school. We went to the quarterfinals of the state championships. We were a real unit. And I was a three-year starter, so you could say I had something to do with that."

   d. "I can get along with Italians and the Irish, Blacks and Puerto Ricans, all kinds of different people."

3. Which of these is the most effective way to market yourself, as you search for a job?

   a. Create a Facebook account on the internet that describes you and your skills and abilities.

   b. Go up and down the streets of your neighborhood, knocking on all your neighbors' doors and telling them what a good construction worker you are.

   c. Print out flyers that list your skills and experiences, and then distribute them to every construction site and post them on every bulletin board in your area.

   d. Look in the newspaper every day and respond to every appropriate want ad you find.

4. You're filling out an online application. It asks if you've ever had trouble with the law. Five years ago, while you were still in high school, you went to juvenile detention. Since then, you've been a straight arrow and haven't gotten even a speeding ticket. How should you handle this issue on the application?

   a. Write in "N/A" in the space provided. The incident was a long time ago, you're as law-abiding as they come now, and so why dredge up the past?

   b. Mention the incident, complete with details on why you were in juvenile detention and for how long.

   c. Look for the part of the application that says Other, and jot down something about having a bit of trouble years ago but that you're a straight shooter now.

   d. Don't mention the incident at all and hope the topic won't come up during the interview; if it does, you can always tell the interviewer you didn't see a place on the application to bring up the subject.

5. The interviewer asks you to talk about yourself. What's the best way to respond?

    a. "Where do I begin? I mean, I've got so many skills and talents I want to tell you about, because I really am a good candidate for this position."

    b. "Well, I've never disagreed with any boss I've ever had. That's one thing. I'm not perfect, though; for instance, I put in too much overtime, which wreaks havoc with my social life."

    c. "That's an interesting question. I mean, a really good one. I'm just not quite sure about what you want me to say. Do you want to hear about when I was in the school band? Or when I had a summer job cutting lawns for Mrs. Lopez down the street?"

    d. "Eventually, I'd like to become a master electrician. I know that's a ways off, but I've got my eye on some classes in it at the local community colleges. While I was reading up about your company, I noticed it's involved with apprenticeship programs with electricians. That's something to shoot for once I get some classroom work under my belt, and it's one of the reasons I'd really like to work here."

6. What should you put on your resume?

    a. Birthday

    b. Education

    c. Social Security Number

    d. Grades from school

7. Your resume should include the following topics, presented in this order: _____.

    a. Objective, skills and qualifications, job experience, and education

    b. Name, education, skills and qualifications, references

    c. Job experience, other experience, skills and achievements, pay requested

    d. Objective, job experience, skills and experience, summary: why you should hire me

8. You've arrived at an interview appointment and are filling out an application. The form asks you about your education, and for some reason, you draw a complete blank. You don't have the information with you. You should _____.

    a. write "Information to Come" across the section of the form asking about your education

    b. tell the receptionist as you hand the form back that you've forgotten some of the education information and that you'll get it to them as soon as possible, and then ask politely where you should send the missing information

    c. make a joke about it during the interview

    d. tell the interviewer that you forgot, but that you'll get that information back to them soon

9. From this list, what is the most important thing you can do to make your job search successful?

    a. Prepare for your job interview.

    b. Prepare your resume.

    c. Never lie or exaggerate during an interview, on your resume, or when filling out your application.

    d. All of the above.

10. You've been searching for a job for awhile now, and not having much luck. You should _____.

    a. move to another state

    b. quit filling out online application forms and just spend time on the ones you mail in

    c. add a bit more detail to your resume about construction skills you don't really have now, but that you know you'll have once you get an actual job

    d. persevere and not get discouraged, and think about whether you are applying to the right kind of companies and for the right kind of positions

Individual Activities

Activity 1: Filling Out Employment Forms

To get a job, you'll have to fill out an application and give the company and the government some personal information. Here are some tips on doing that. After you read through the tips, complete the I-9 Form *(Figure 1–1)* and the employment application form *(Figure 1–2)*. The U.S. Bureau of Citizenship and Immigration Services (a division of the Department of Homeland Security) requires the I-9 Form for anyone (citizens and noncitizens) hired for employment in the United States.

Tips for Filling Out the Employment Application and I-9 Forms

1. **Take your time.** Always read through the whole form before you begin filling it out. (Note that on some forms, such as the I-9, your employer fills out part of the form for you.)

2. **Ask for help.** If there is something about the form you don't understand, don't be shy about asking for help.

3. **Follow directions exactly.** On some forms you must write your first name first. On others, you must write your last name first. Almost all employment applications require you to list your experience starting with your most recent employer.

4. **Look before you write.** Take a look at how much space you have to write in. Some forms don't give you very much space, so you may have to print a bit smaller than usual.

5. **Think before you write.** You must complete printed forms in ink. You'll have to cross out any mistakes, and there won't be much room to squeeze in corrections. Pay special attention to boxes that you must check, especially any that are next to the words *Yes* and *No*. Be sure to check the right box. (Note that some companies may allow you to complete an application online.)

6. **Print clearly and neatly.** Think about the person who has to read your application. Make it as easy to read as possible. The person who reads your application may one day be your boss.

7. **Don't leave blanks.** To avoid leaving blanks on an application, bring a copy of your resume with you. If you don't have a resume, bring along a folder or a piece of paper with information about you, your education, and your work experience. Here are some examples of information you will need:

   • The names and addresses of any schools you attended and when you attended them.

   • Information about military service.

   • Former addresses (some companies want this information if you've lived at your current address for less than five years).

   • Names, addresses, and phone numbers of former employers.

   • Names, addresses, and phone numbers of references.

8. **Be truthful.** Many forms related to getting jobs include a penalty if you don't tell the truth about your experience or background.

9. **Review the form.** Check over what you have written to make sure it is accurate and that you have not left anything out.

10. **Sign.** Most employment forms require a signature and date. Don't forget this last step.

Figure 1–1. I-9 form (1 of 5)

OMB No. 1615-0047; Expires 08/31/12

Department of Homeland Security
U.S. Citizenship and Immigration Services

Form I-9, Employment Eligibility Verification

Instructions
Read all instructions carefully before completing this form.

Anti-Discrimination Notice. It is illegal to discriminate against any individual (other than an alien not authorized to work in the United States) in hiring, discharging, or recruiting or referring for a fee because of that individual's national origin or citizenship status. It is illegal to discriminate against work-authorized individuals. Employers **CANNOT** specify which document(s) they will accept from an employee. The refusal to hire an individual because the documents presented have a future expiration date may also constitute illegal discrimination. For more information, call the Office of Special Counsel for Immigration Related Unfair Employment Practices at 1-800-255-8155.

What Is the Purpose of This Form?

The purpose of this form is to document that each new employee (both citizen and noncitizen) hired after November 6, 1986, is authorized to work in the United States.

When Should Form I-9 Be Used?

All employees (citizens and noncitizens) hired after November 6, 1986, and working in the United States must complete Form I-9.

Filling Out Form I-9

Section 1, Employee

This part of the form must be completed no later than the time of hire, which is the actual beginning of employment. Providing the Social Security Number is voluntary, except for employees hired by employers participating in the USCIS Electronic Employment Eligibility Verification Program (E-Verify). **The employer is responsible for ensuring that Section 1 is timely and properly completed.**

Noncitizen nationals of the United States are persons born in American Samoa, certain former citizens of the former Trust Territory of the Pacific Islands, and certain children of noncitizen nationals born abroad.

Employers should note the work authorization expiration date (if any) shown in **Section 1**. For employees who indicate an employment authorization expiration date in **Section 1**, employers are required to reverify employment authorization for employment on or before the date shown. Note that some employees may leave the expiration date blank if they are aliens whose work authorization does not expire (e.g., asylees, refugees, certain citizens of the Federated States of Micronesia or the Republic of the Marshall Islands). For such employees, reverification does not apply unless they choose to present

in Section 2 evidence of employment authorization that contains an expiration date (e.g., Employment Authorization Document (Form I-766)).

Preparer/Translator Certification

The Preparer/Translator Certification must be completed if **Section 1** is prepared by a person other than the employee. A preparer/translator may be used only when the employee is unable to complete **Section 1** on his or her own. However, the employee must still sign **Section 1** personally.

Section 2, Employer

For the purpose of completing this form, the term "employer" means all employers including those recruiters and referrers for a fee who are agricultural associations, agricultural employers, or farm labor contractors. Employers must complete **Section 2** by examining evidence of identity and employment authorization within three business days of the date employment begins. However, if an employer hires an individual for less than three business days, **Section 2** must be completed at the time employment begins. Employers cannot specify which document(s) listed on the last page of Form I-9 employees present to establish identity and employment authorization. Employees may present any List A document **OR** a combination of a List B and a List C document.

If an employee is unable to present a required document (or documents), the employee must present an acceptable receipt in lieu of a document listed on the last page of this form. Receipts showing that a person has applied for an initial grant of employment authorization, or for renewal of employment authorization, are not acceptable. Employees must present receipts within three business days of the date employment begins and must present valid replacement documents within 90 days or other specified time.

Employers must record in Section 2:

1. Document title;
2. Issuing authority;
3. Document number;
4. Expiration date, if any; and
5. The date employment begins.

Employers must sign and date the certification in **Section 2**. Employees must present original documents. Employers may, but are not required to, photocopy the document(s) presented. If photocopies are made, they must be made for all new hires. Photocopies may only be used for the verification process and must be retained with Form I-9. **Employers are still responsible for completing and retaining Form I-9.**

Form I-9 (Rev. 08/07/09) Y

Figure 1–1. I-9 form (2 of 5)

For more detailed information, you may refer to the *USCIS Handbook for Employers* (Form M-274). You may obtain the handbook using the contact information found under the header "USCIS Forms and Information."

Section 3, Updating and Reverification

Employers must complete **Section 3** when updating and/or reverifying Form I-9. Employers must reverify employment authorization of their employees on or before the work authorization expiration date recorded in **Section 1** (if any). Employers **CANNOT** specify which document(s) they will accept from an employee.

A. If an employee's name has changed at the time this form is being updated/reverified, complete Block A.

B. If an employee is rehired within three years of the date this form was originally completed and the employee is still authorized to be employed on the same basis as previously indicated on this form (updating), complete Block B and the signature block.

C. If an employee is rehired within three years of the date this form was originally completed and the employee's work authorization has expired **or** if a current employee's work authorization is about to expire (reverification), complete Block B; and:

    1. Examine any document that reflects the employee is authorized to work in the United States (see List A **or** C);

    2. Record the document title, document number, and expiration date (if any) in Block C; and

    3. Complete the signature block.

Note that for reverification purposes, employers have the option of completing a new Form I-9 instead of completing **Section 3.**

What Is the Filing Fee?

There is no associated filing fee for completing Form I-9. This form is not filed with USCIS or any government agency. Form I-9 must be retained by the employer and made available for inspection by U.S. Government officials as specified in the Privacy Act Notice below.

USCIS Forms and Information

To order USCIS forms, you can download them from our website at www.uscis.gov/forms or call our toll-free number at 1-800-870-3676. You can obtain information about Form I-9 from our website at www.uscis.gov or by calling 1-888-464-4218.

Information about E-Verify, a free and voluntary program that allows participating employers to electronically verify the employment eligibility of their newly hired employees, can be obtained from our website at www.uscis.gov/e-verify or by calling 1-888-464-4218.

General information on immigration laws, regulations, and procedures can be obtained by telephoning our National Customer Service Center at 1-800-375-5283 or visiting our Internet website at www.uscis.gov.

Photocopying and Retaining Form I-9

A blank Form I-9 may be reproduced, provided both sides are copied. The Instructions must be available to all employees completing this form. Employers must retain completed Form I-9s for three years after the date of hire or one year after the date employment ends, whichever is later.

Form I-9 may be signed and retained electronically, as authorized in Department of Homeland Security regulations at 8 CFR 274a.2.

Privacy Act Notice

The authority for collecting this information is the Immigration Reform and Control Act of 1986, Pub. L. 99-603 (8 USC 1324a).

This information is for employers to verify the eligibility of individuals for employment to preclude the unlawful hiring, or recruiting or referring for a fee, of aliens who are not authorized to work in the United States.

This information will be used by employers as a record of their basis for determining eligibility of an employee to work in the United States. The form will be kept by the employer and made available for inspection by authorized officials of the Department of Homeland Security, Department of Labor, and Office of Special Counsel for Immigration-Related Unfair Employment Practices.

Submission of the information required in this form is voluntary. However, an individual may not begin employment unless this form is completed, since employers are subject to civil or criminal penalties if they do not comply with the Immigration Reform and Control Act of 1986.

EMPLOYERS MUST RETAIN COMPLETED FORM I-9
DO NOT MAIL COMPLETED FORM I-9 TO ICE OR USCIS

Form I-9 (Rev. 08/07/09) Y Page 2

Figure 1–1. I-9 form (3 of 5)

Paperwork Reduction Act

An agency may not conduct or sponsor an information collection and a person is not required to respond to a collection of information unless it displays a currently valid OMB control number. The public reporting burden for this collection of information is estimated at 12 minutes per response, including the time for reviewing instructions and completing and submitting the form. Send comments regarding this burden estimate or any other aspect of this collection of information, including suggestions for reducing this burden, to: U.S. Citizenship and Immigration Services, Regulatory Management Division, 111 Massachusetts Avenue, N.W., 3rd Floor, Suite 3008, Washington, DC 20529-2210. OMB No. 1615-0047. **Do not mail your completed Form I-9 to this address.**

Form I-9 (Rev. 08/07/09) Y Page 3

Figure 1–1. I-9 form (4 of 5)

OMB No. 1615-0047; Expires 08/31/12

Department of Homeland Security
U.S. Citizenship and Immigration Services

**Form I-9, Employment
Eligibility Verification**

Read instructions carefully before completing this form. The instructions must be available during completion of this form.

ANTI-DISCRIMINATION NOTICE: It is illegal to discriminate against work-authorized individuals. Employers CANNOT specify which document(s) they will accept from an employee. The refusal to hire an individual because the documents have a future expiration date may also constitute illegal discrimination.

Section 1. Employee Information and Verification *(To be completed and signed by employee at the time employment begins.)*

Print Name: Last	First	Middle Initial	Maiden Name

Address *(Street Name and Number)*		Apt. #	Date of Birth *(month/day/year)*

City	State	Zip Code	Social Security #

I am aware that federal law provides for imprisonment and/or fines for false statements or use of false documents in connection with the completion of this form.

I attest, under penalty of perjury, that I am (check one of the following):

☐ A citizen of the United States

☐ A noncitizen national of the United States (see instructions)

☐ A lawful permanent resident (Alien #) _____

☐ An alien authorized to work (Alien # or Admission #) _____
until (expiration date, if applicable - *month/day/year*) _____

Employee's Signature   Date *(month/day/year)*

Preparer and/or Translator Certification *(To be completed and signed if Section 1 is prepared by a person other than the employee.)* I attest, under penalty of perjury, that I have assisted in the completion of this form and that to the best of my knowledge the information is true and correct.

Preparer's/Translator's Signature	Print Name

Address *(Street Name and Number, City, State, Zip Code)*	Date *(month/day/year)*

Section 2. Employer Review and Verification *(To be completed and signed by employer. Examine one document from List A OR examine one document from List B and one from List C, as listed on the reverse of this form, and record the title, number, and expiration date, if any, of the document(s).)*

	List A	OR	List B	AND	List C
Document title:	_____		_____		_____
Issuing authority:	_____		_____		_____
Document #:	_____		_____		_____
Expiration Date *(if any)*:	_____		_____		_____
Document #:	_____				
Expiration Date *(if any)*:	_____				

CERTIFICATION: I attest, under penalty of perjury, that I have examined the document(s) presented by the above-named employee, that the above-listed document(s) appear to be genuine and to relate to the employee named, that the employee began employment on *(month/day/year)* _____ and that to the best of my knowledge the employee is authorized to work in the United States. (State employment agencies may omit the date the employee began employment.)

Signature of Employer or Authorized Representative	Print Name	Title

Business or Organization Name and Address *(Street Name and Number, City, State, Zip Code)*	Date *(month/day/year)*

Section 3. Updating and Reverification *(To be completed and signed by employer.)*

A. New Name *(if applicable)*	B. Date of Rehire *(month/day/year) (if applicable)*

C. If employee's previous grant of work authorization has expired, provide the information below for the document that establishes current employment authorization.

Document Title: _____	Document #: _____	Expiration Date *(if any)*: _____

I attest, under penalty of perjury, that to the best of my knowledge, this employee is authorized to work in the United States, and if the employee presented document(s), the document(s) I have examined appear to be genuine and to relate to the individual.

Signature of Employer or Authorized Representative	Date *(month/day/year)*

Form I-9 (Rev. 08/07/09) Y Page 4

Figure 1–1. I-9 form (5 of 5)

LISTS OF ACCEPTABLE DOCUMENTS
All documents must be unexpired

LIST A		LIST B		LIST C
Documents that Establish Both Identity and Employment Authorization	**OR**	**Documents that Establish Identity**	**AND**	**Documents that Establish Employment Authorization**

LIST A	LIST B	LIST C
1. U.S. Passport or U.S. Passport Card	1. Driver's license or ID card issued by a State or outlying possession of the United States provided it contains a photograph or information such as name, date of birth, gender, height, eye color, and address	1. Social Security Account Number card other than one that specifies on the face that the issuance of the card does not authorize employment in the United States
2. Permanent Resident Card or Alien Registration Receipt Card (Form I-551)	2. ID card issued by federal, state or local government agencies or entities, provided it contains a photograph or information such as name, date of birth, gender, height, eye color, and address	2. Certification of Birth Abroad issued by the Department of State (Form FS-545)
3. Foreign passport that contains a temporary I-551 stamp or temporary I-551 printed notation on a machine-readable immigrant visa		3. Certification of Report of Birth issued by the Department of State (Form DS-1350)
4. Employment Authorization Document that contains a photograph (Form I-766)	3. School ID card with a photograph	4. Original or certified copy of birth certificate issued by a State, county, municipal authority, or territory of the United States bearing an official seal
	4. Voter's registration card	
5. In the case of a nonimmigrant alien authorized to work for a specific employer incident to status, a foreign passport with Form I-94 or Form I-94A bearing the same name as the passport and containing an endorsement of the alien's nonimmigrant status, as long as the period of endorsement has not yet expired and the proposed employment is not in conflict with any restrictions or limitations identified on the form	5. U.S. Military card or draft record	
	6. Military dependent's ID card	5. Native American tribal document
	7. U.S. Coast Guard Merchant Mariner Card	
	8. Native American tribal document	6. U.S. Citizen ID Card (Form I-197)
	9. Driver's license issued by a Canadian government authority	
	For persons under age 18 who are unable to present a document listed above:	7. Identification Card for Use of Resident Citizen in the United States (Form I-179)
6. Passport from the Federated States of Micronesia (FSM) or the Republic of the Marshall Islands (RMI) with Form I-94 or Form I-94A indicating nonimmigrant admission under the Compact of Free Association Between the United States and the FSM or RMI	10. School record or report card	8. Employment authorization document issued by the Department of Homeland Security
	11. Clinic, doctor, or hospital record	
	12. Day-care or nursery school record	

Illustrations of many of these documents appear in Part 8 of the Handbook for Employers (M-274)

Form I-9 (Rev. 08/07/09) Y Page 5

Figure 1–2. Employment Application (1 of 4)

NabholzConstruction

Building Integrity

612 Garland Street ■ Conway, Arkansas 72032 ■ Ph 501-327-7781 ■ Fax 501-327-8231 ■ www.nabholz.com

NOTICE TO ALL APPLICANTS

RE: DRUG & ALCOHOL ABUSE POLICY

Employers have certain necessary and valid conditions of employment which must be met to ensure a safe workplace. Due to the nature of our operation, employees must be able to exercise good judgment, react properly in unexpected situations and perform tasks efficiently and safely.

We care about the safety and health of our employees, including you, if you are employed with us. Therefore, as a condition for employment, ALL applicants must be drug screened before they can be considered for employment.

In our continued effort to maintain a safe and healthy workplace, as required by Federal Law, we plan to include both scheduled and unscheduled drug/alcohol testing of employees, as well as programs with which to assist our employees who may have or develop these substance abuse problems. Our substance abuse policy and consent and release form are included in this application for employment.

> If you use illegal drugs or if you excessively drink alcoholic beverages, you may NOT want to complete this application; rather, you may want to seek employment elsewhere.

Thank you for your interest in working for Nabholz.

The Nabholz Companies
Employee Services Department

"Nabholz...we will be the best. We care about people."

ACTIVE EMPLOYEE CERTIFICATE AGREEMENT
This certificate becomes part of the active employee's personnel file.

NABHOLZ COMPANIES

I do hereby certify that I have received and read the NABHOLZ COMPANIES substance abuse and testing policy and have had the drug-free workplace program explained to me. I understand that if my performance indicates it is necessary, I will submit to a drug and/or alcohol test. I also understand that failure to comply with a drug and/or alcohol testing request or a positive, confirmed result for the illegal use of drugs and/or alcohol may lead to discipline up to and including termination of employment and/or loss of workers compensation benefits, pursuant to Nabholz Company Policy.

_____
Employee Signature

_____
Employee Name (Print)

_____
Date

Source: Nabholz Construction Corporation, Conway, AR

Figure 1–2. Employment Application (2 of 4)

Building Integrity

612 Garland Street ▪ Conway, Arkansas 72032 ▪ Ph 501-327-7781 ▪ Fax 501-327-8231 ▪ www.nabholz.com

APPLICATION FOR EMPLOYMENT
AN EQUAL EMPLOYMENT OPPORTUNITY EMPLOYER M/F

Note: This application is valid for 60 days. If you wish to be considered for employment after this 60-day period, a new application must be completed.

PERSONAL INFORMATION   DATE:_____

Name:_____ SS#:_____

Present
Address:_____
                Street   City   State   Zip

Permanent
Address:_____
                Street   City   State   Zip

Phone No:_____ Are you 18 years or older? YES____ NO____

EMPLOYMENT DESIRED

Position:_____ Date You Can Start_____ Salary Desired_____

Are You Employed Now?_____ If So, May We Inquire of Your Present Employer?_____

Ever Applied to
This Company Before?_____ If so, where?_____ When?_____

EDUCATION	Name & Location of School	No. of years Attended	Did You Graduate?	Subjects Studied
Grammer				
High School				
College				
Trade or Business				

GENERAL

Subjects of special study or research work:_____
Military or Naval Service: _____ Rank:_____
Present Membership in National Guard or Reserves:_____

FORMER EMPLOYERS   (List last four, starting with last one)

Date/Month/Year	Name &Address	Supervisor	Salary	Position	Reason for Leaving
1. to					
2. to					
3. to					
4. to					

Figure 1–2. Employment Application (3 of 4)

REFERENCES
(Give the names of three persons not related to you, whom you have known at least one year.)

Name	Address	Business	Year Acquainted

1._____

2._____

3._____

In case of Emergency, notify:_____

                                        Name   Address

Phone:_____

Have you ever been convicted of a crime, other than minor traffic offenses:_____

If so, please explain_____

Note: A prior conviction will not necessarily bar you from employment; however the type of conviction and when it occurred will be considered.

CERTIFICATION

"I certify that the information in this application is true and understand that misrepresentations of false or omitted facts may result in my termination, regardless of the time of discovery by the company. I also understand that, if hired, my employment is for no definite period and may be terminated at any time without written notice and that, absent a written contract signed by the President of the company, I will remain an at-will employee and can be terminated at any time without any notice.

I authorize investigation of the statements contained herein and the references listed above to give you any and all information concerning my previous employment and any pertinent information such references may have, personal or otherwise, and release all parties from all liability for any damage that may result from furnishing same to you.

I understand that if the company decides to engage an investigative consumer reporting agency to report on my credit and personal history, the company will provide me, at my request, with the name and address of the agency so that I can obtain from them the nature and substance of the information contained in the report.

_____

    Date   Signature

DO NOT WRITE BELOW THIS LINE

Interviewed By:_____   Date:_____

Hire: Yes_____ No_____   Position:_____

Department: _____ Salary:_____

Date Reporting to Work:_____

Approved: 1._____ 2._____ 3._____

Figure 1–2. Employment Application (4 of 4)

Building Integrity
612 Garland Street ■ Conway, Arkansas 72032 ■ Ph 501-327-7781 ■ Fax 501-327-8231 ■ www.nabholz.com

It is the policy of the Nabholz Construction Corporation, Inc. to ensure and maintain a working environment free of harassment, intimidation, and coercion at all sites, and in all facilities, at which our employees are assigned to work. Specific attention will be given to ensure that minorities and women are provided with a work environment free of harassment, intimidation and coercion at all times. Any harassment, intimidation, or coercion observed by any employee should be reported immediately to your supervisor or the Company EEO Officer. This policy will be rigidly adhered to by all personnel of the Nabholz Construction Corporation.

It is the policy of this organization to provide equal employment opportunity to all qualified applicants for employment without regard to race, color, religion, national origin, sex, age, veteran status or disability.

I have read the above Company Policy and have been provided other information, related to Company and Affirmative Action Procedures. As a condition of my employment, I hereby agree to comply with the above policy and report any violations of this policy to the Company EEO Officer, and/or my supervisor.

_____   _____
Employee Signature   Date

To help us comply with Federal/State Equal Employment Opportunity record keeping, reporting and other legal requirements, please answer the questions below. **This form is confidential and will be maintained separately from your application form.**

NAME:
(print) _____
　　　　LAST　　　　　　　　　　　FIRST　　　　　　　　MIDDLE

ADDRESS: _____
　　　　　Street　　　　　　　　　　City　　　　　　　State/Zip Code

TELEPHONE: (__) _____ BIRTHDATE _____ AGE _____

SEX: ( ) Male   ( ) Female

MARITAL STATUS: ( ) Single   ( ) Married   ( ) Divorced   ( ) Widowed

RACE/ETHNIC GROUP: ( ) White   ( ) African-American   ( ) Hispanic
　　　　　( ) American Indian/Alaskan Native   ( ) Asian/Pacific Islander

Are you a Disabled Veteran? ( ) Yes   ( ) No   If yes what is your VA Disability Rating? _____ %
A person entitled to disability compensation under laws administered by the Veterans Administration for disability rated at 30% or more, or a person whose discharge or release from active duty was for a disability incurred or aggravated in the line of duty.

Are you a Vietnam Veteran? ( ) Yes   ( ) No
A person who served on active duty for a period of more than 180 days any part of which occurred between 8/5/64 and 5/7/75, and was discharged or released there from with other than a dishonorable discharge or for a service connected disability.

Activity 2: Writing a Resume

Your resume is one of the most important work documents you will write. In many cases, the appearance and content of your resume will determine whether you will have a face-to-face interview. In this activity, you will collect information to write a resume for yourself. Refer to the sample resume on the following page as a suggestion for how your final document should look.

Questions to Ask Yourself	Statements or Information to Include on your Resume
Name, address, phone number, email address	
1. What type of job are you hoping to get?	
2. Who have you worked for already? Include company names, addresses, and phone numbers.	
3. What types of work have you done?	
4. What types of skills do you have?	
5. What type of education have you received? Include your apprenticeship training plus any certification programs you have completed.	

Your Name
Address
Phone number (home and cell, as applicable)
Email address (if applicable)

EMPLOYMENT OBJECTIVE

I am seeking a job in construction where I can use my broad carpentry and woodworking skills. My eventual goal is to become a general construction supervisor, through exposure to the entire construction process as I progress in my carpentry career.

SKILLS AND QUALIFICATIONS

- Practical knowledge of carpentry hand and power tools, measuring tools, and woodworking tools.
- Qualified to read residential and commercial blueprints.
- Experience with both stick and modular frames, including steel frames.
- Working knowledge of roofing and residential plumbing.
- Experienced in trench work and excavation, including the installation of sewer, water, and storm drain pipes.
- Trained and certified to use powder-actuated tools.
- Certified in scaffold building.

JOB EXPERIENCE

May 2006 to present. Journeyman carpenter. LJL Construction, 123 Hammer Heights Rd., Fairfax, Va. 800-222-2222. Built custom cabinets, closet systems; framed single-family homes. Responsible for, among other projects, the remodeling of the Regan Community Center, Annandale, Va.

May 2004 to May 2006. Carpenter. Hammer Company, 456 Lathe Lane, Philadelphia, Pa. 900-456-4566. Built closets and storage systems for commercial storage units. Worked on more than 75 different projects, both remodeling and new construction, in the Philadelphia region.

June 2002 to May 2004. Carpenter. Blue Ridge Builders, 789 Mountain Way, Roanoke, Va. 888-123-4567. Framed condo units, built fences, and raised wood walkways.

April 1999 to June 2002. Construction Laborer. Barton Arms Associates, 111 Buena Vista Blvd., Winston-Salem, N.C. 800-555-1212. Installed utility lines, erected and disassembled scaffolding and other temporary structures. Performed layout, rough framing, and outside finishing. Responsible for materials ordering and delivery.

CONSTRUCTION EDUCATION

- Powder-actuated tools certification program. All States Community College, Fairfax, Va. Certificate, 2008.
- Professional carpentry certification program. ABC Training Corp., Philadelphia, Pa. Certificate, 2006.
- NCCER journey-level assessments in form and frame carpentry, 2004.
- Apprenticeship. Carpentry. MK Builders Inc., Winston-Salem, N.C. Certificate, 2002.
- High school diploma, W.P. Mayhew High School, Richmond, Va., 1998.

Activity 3: Practicing Your Pitch

In this exercise, you'll come up with a way to sell yourself (known as the pitch) to an interviewer. Then you'll practice that pitch in front of a group of your classmates. Break into teams of four. Each person will then come up with a quick pitch to sell him or herself, and will deliver it to his or her teammates. Then the teammates will discuss how effective the pitch was and what could be done to improve it. Continue until all teammates have had a turn.

Professionalism: Keeping Your Job

"Work eight hours and sleep eight hours,
   and make sure that they are not the same hours."

– T. Boone Pickens,
oil executive

Introduction

Imagine that you have hired two people, a carpenter and a stonemason, to work on your house. You have a limited budget, and the work needs to be done quickly. The carpenter shows up to work on time, treats you with respect, takes pride in the work, and in general is mature and businesslike. The carpentry work gets done fast and well. On the other hand, the stonemason often shows up late and leaves early, is cranky around you and other workers, leaves tools lying around, and takes much longer to finish the job than expected. The job is completed poorly. Which one of these workers would you want to hire again? Which one would you recommend to your friends?

Just as you expect the people you hire to act and work in certain ways, your employer expects you to act and work the right way. This means acting professionally: being dependable, organized, honest, prepared, and respectful. Professionals always show up on time and when expected, and do not leave early. They do not stretch out their lunch breaks or cut corners.

Acting professionally is not only necessary to keep your job, it's the key to advancing your career. Trust-worthy professionals are valued by their employers, and they get a good reputation for working hard, treating everyone with respect, and being honest and dependable. These are the workers companies want to hire and retain. If you consistently demonstrate that you have these qualities, companies might end up competing for your services. That means more prestige—and often more money—for you.

In this module, you will learn what employers expect from their employees, and you will see how acting professionally pays off in many ways.

What are employers looking for?

When you are just starting out in construction, you probably won't have as many skills or as much experience as many of your co-workers. Most companies and supervisors do not expect someone just starting out to be an expert. They do expect you to work hard, be dependable and honest, and treat your bosses, co-workers, and the company with respect. The following are some qualities that employers want to see in their employees:

Dependable: Dependable workers can be trusted to get the job done correctly and promptly. They show up for work every day and are always on time. Dependable workers do what they promise, and are therefore trusted by their bosses.

Work-oriented: To be work-oriented means enjoying your job and taking pride in your work. Because their work is important to them, work-oriented employees believe in giving a day's work for a day's pay. Supervisors respect their work-oriented employees, and will not have much patience with slackers.

Organized: Organized workers have a plan for what they want to do every day. They keep their tools and work areas neat and clean. Being organized and prepared helps to keep the worksite safe and the project on schedule. If half the day is spent looking for a tool or cleaning up after the previous shift, not much new work gets done. Organized workers do not leave their tools lying around where they might trip co-workers or fall into machinery. They do not leave debris in the work area and are careful to properly store or dispose of flammable materials.

Technically qualified: Becoming technically qualified does not happen by accident. Technically proficient workers get that way because they keep up with advances in the industry and are always looking to expand their skills. Technically qualified workers know how to operate machinery safely, and they learn what tools and techniques are needed for a job before they start.

Flexible: The best workers are willing to learn new tasks and try new ideas. Flexible workers pitch in to help no matter what the task. Even boring jobs can be learning opportunities. Supervisors take note of those workers most willing to help out, and they think more highly of them than of those workers who say, "This isn't my job."

Honest: Honest workers are not only personally honest, they value honesty in others. Honest workers call in sick only when they are actually ill, take good care of their employer's property and tools, and do not borrow those items for personal use. Honest workers leave work early only when they have arranged with their supervisor to do so. If they are struggling with a problem, honest workers seek the help of a supervisor and do not just ignore it or try to fix it just enough to get by.

Prepared: The best workers know which tools they are responsible for and keep those tools in good working order. Professional workers know that maintaining their tools is essential to both the job and safety. They check their tools for frayed cords, dull edges, missing or jammed parts, or anything else that might cause an accident. Prepared workers also know how to follow manufacturers' guidelines for using, maintaining, and repairing tools. Supervisors know that those workers who keep their tools prepared for use are workers who are generally well-prepared.

Respectful of the rules: Professionals realize rules are there for a reason, and they always follow those rules. They know that construction sites are full of hazards that can cause injury or even death, so they always wear the right safety gear and follow all safety procedures. Professionals also follow company rules designed to keep the project on time and within budget, because they know that their raises and promotions depend on how well their company performs.

Respectful of the company: Professionals recognize that they represent their company even when they are not on the worksite. They are careful about what they say when they talk about their company. Employers do not want workers who speak negatively about the company. Respecting the company also means taking care of company equipment and using work time only for company-related projects.

Well-groomed and appropriately dressed: Professional workers realize the importance of dressing appropriately for the job. Well-groomed does not mean having to get a manicure or to shave every single day; it means being neat and clean. At the worksite, sloppy dress can be a safety hazard. Long, messy hair can get caught in machinery and untied shoelaces can cause falls. Sloppy dress may also get in the way of wearing special equipment required for certain jobs. Employers feel that workers who take pride in their appearance are apt to take pride in their work.

Did You Know ...?

... that two-thirds of workers who call in sick at the last moment do so for reasons other than physical illness? That more than two-thirds of employers can find patterns in unscheduled absences? Unscheduled absenteeism costs U.S. companies billions of dollars a year in direct costs related to paychecks and in indirect costs from lower productivity and the effect unscheduled absenteeism has on co-workers' morale. A 2007 survey found that personal illness made up only 34 percent of unscheduled absences, while 66 percent were from other reasons such as family issues, personal needs, stress, and what the study called an "entitlement mentality." There may be times you feel justified to call in sick for reasons other than physical illness, but these should be rare. Some reasons are not valid at all: for instance, feeling entitled to a day off because you've put in so much overtime recently.

Two keys to success: Attendance and punctuality

To make a profit and stay in business, construction companies operate under tight schedules. Keeping to those schedules requires reliable workers. If a worker is late or does not show up, immediate adjustments become necessary, which add rapidly to costs. An unreliable crew member affects the entire team.

When supervisors talk about the most common problems they face on the job, two in particular come up: lateness and absenteeism. Workers with reputations for constantly being late or not showing up as scheduled will not get very far in the construction profession, which depends a lot on its employees' reliability. Here are some suggestions to help you achieve an excellent record of attendance and punctuality.

Think about what would happen if all workers were late or frequently absent. Empathy is when you think about the consequences of your actions on other people, and when you can put yourself in another person's place and understand how they feel. It is one of the most critical skills to have at the workplace. Put yourself in your supervisor's shoes: how could you get the job done on budget and on time if workers were always late or just didn't show up? Would you worry about your workers' safety if there weren't enough people around to do the job properly?

Think about your co-workers. It is also important to have empathy for your co-workers. When you are late or absent, they are the ones who have to pick up the slack. It is not fair to them if you are always late or calling in sick when you are not actually ill. Good attendance and being on time are very important factors in workplace safety. Employees who have to work longer hours or do more tasks to fill in for a perpetually late or absent worker might get tired sooner, which can lead to accidents.

Understand when it is OK to call in sick. Call in sick only when you are actually ill. All companies have rules on how their employees can use sick leave. Some companies may allow the use of sick leave for family emergencies, or they may give employees personal leave days to use in case of emergencies. You must learn and follow your company's sick leave policy. If your company has a human resources department, find out how much leave you have and how you can use it. Whether it is sick leave or personal leave, use it honestly. Do not take leave except for a good reason.

Notify your supervisor as soon as possible if you are running late or must miss work. Call your supervisor as soon as you know that you will be late or absent. It is best to call as close as possible to the start of your shift. Unless you are unable to, make the call yourself; do not ask someone else to call in for you.

Keep your supervisor posted if you have to be out for more than a day. Your absence affects scheduling and work assignments, so let your boss know when you will be back on the job. If you know you'll have to miss work in the future due to a non-illness reason (for instance, serving on a jury), let your boss know as soon as possible what days you'll be absent so that necessary adjustments can be made in advance.

Give yourself enough time to get to work. Part of being well-organized is paying attention to conditions that can affect your commute before you set out for work. If you are driving, turn on the radio (or go online if you have access to a computer) to find out the road conditions on the route you normally take. Allow more time to get to work if traffic or weather conditions are causing delays, and find a different route if your usual one is blocked. If traffic does not turn out to be bad and you end up showing up to work early, you'll look good for arriving a few minutes before everyone else!

Go to your supervisor immediately when you are late. Do not try to sneak onto the jobsite hoping that nobody notices. Report to your boss, explain why you are late, and apologize. Take responsibility for your actions.

Remember, good attendance means more than showing up for work regularly and on time. It also means using only the allotted amount of time for your meals or breaks, and it means working until the end of your scheduled shift. If suddenly you have to leave work early, you must still tell your boss before you go, and you must have a good reason (for instance, to attend to an emergency).

Get enough rest. In most construction jobs, you will start your workday very early; therefore, you should not stay up late on work nights. When you get enough rest, you will be alert and able to do your job to the best of your abilities. Do not party on nights before a workday.

A good attitude: The right start

During your schooling and training, you will learn many technical skills. You will fine-tune these skills on the jobsite working with skilled craftworkers. Technical skill is very important to your success in the construction industry, but your attitude is almost as important. Would you rather work with someone who complains all the time and does not treat you with respect, or someone who is agreeable, enjoys working, and treats you courteously? All else being equal, most employers would rather hire an upbeat employee. Workers with a good attitude tend to be the most productive, and they have a positive influence on the entire work crew.

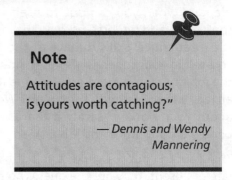

Note

Attitudes are contagious; is yours worth catching?"

— *Dennis and Wendy Mannering*

How can you keep your attitude positive? It's not hard to do if you keep a few things in mind:

Be pleasant. The construction industry offers jobs that reward hard work and skill. When you feel good about yourself, it's easy to feel good about others, too. Being pleasant to those around you shows you are approachable and can talk to others, whether they are co-workers or clients.

Take pride in your work and in your tools. No matter what you do on the construction site, take pride in the skills that you have been working so hard to develop. Take pride in the fact you are working hard, period. Let your pride show in every task you do. Do not let pride slip into arrogance, though.

Cooperate. Lend a hand when someone needs help. "That's not my job" shouldn't be in your vocabulary. If you are given a task you do not like, think of it as a way to gain valuable experience or as an opportunity to prove your willingness to help. Any task can be a chance to learn something. When your supervisor asks you to do something you normally don't do, do it without complaining.

Take initiative. When you report to the jobsite, start working on time—do not wait for someone to tell you to start work. Take charge of your job. If you finish a task and have not been told what to do next, ask your boss; part of your supervisor's job is to make sure everyone stays busy.

Accept responsibility. It's inevitable that you will make mistakes. Everyone does. It is important to accept responsibility for them. Do not try to blame someone else for mistakes you make. You must never try to cover up a mistake; doing so could hide a problem that might result in workers being injured or killed. Accepting responsibility is more than just saying "I'm sorry." Instead, explain what happened calmly and clearly. Then work to correct the problem and help get the project back on track.

Have a sense of humor. It's good to have a sense of humor, and it makes the workday more enjoyable. But having a sense of humor does not mean you should become the workplace clown or tell off-color jokes (which many people consider to be sexual harassment). Before you share a joke with your co-workers or supervisor, make sure the joke will not be taken the wrong way. Having a sense of humor means not taking yourself too seriously and being able to laugh at yourself.

Listen and keep an open mind. Imagine you've come up with what you think is a great idea, but your boss or co-workers shoot it down immediately by saying, "What a waste of time! That will never work!" That wouldn't make you feel too good, would it? You want people to listen to you with an open mind, so return the favor. When you are willing to listen, others will talk to you and you are apt to learn something new and interesting.

Be customer-centered. You may not think you have customers, but you do. Everyone who is immediately affected by your work is your customer—the company's client, the owner of the company, your supervisor, even your co-workers. During the workday, you are surrounded by your customers, which is why you must try to deliver your best effort at all times. Keep in mind that you are a representative of your company, even when you are not at work. Do not do or say anything that would embarrass your supervisor or the company.

Regard your job as a learning experience. In construction, the new hires often end up doing what nobody else wants to do—sweeping up the jobsite, cleaning machinery, putting tools away, and loading or unloading materials. Sweeping the floor is probably not what you had in mind when you applied for the job, but everyone has to start somewhere. Keep in mind that many of your co-workers probably started the same way. Do all of your assigned tasks willingly and do them well. Your supervisor will see that you have a good work ethic, and soon other new hires will be doing those tasks.

A word (or two) of advice

How and when to make suggestions. Thinking of new ideas or processes is a great way to get more satisfaction out of your job and to convey a positive attitude. Maybe you've thought of a way to improve how a task is done, and you want to suggest it. Before you bring up your proposal with your supervisor, make sure you can clearly and quickly explain why you think your idea is better. Then make your suggestion when your boss is not busy, preoccupied, or under a lot of pressure. Explain what the issue or problem is and then how your idea or solution could improve things.

Using titles. When in doubt, it is better to be formal than informal when you are new to a work crew. That means using titles such as Mr. or Ms., especially when you are talking to the company owners or your supervisors. In today's workplace, almost everyone is on a first-name basis, but it cannot hurt taking the more formal route when addressing others until they let you know it is OK to use first names.

Workplace culture

A workplace culture develops as a result of the day-to-day actions of owners, supervisors, and co-workers. In some workplaces, people might be formal and use titles. Other workplaces may be more informal and everyone uses first names, even workers, when talking to their bosses. A company's culture is reflected in its values, goals, and climate. When deciding where you will work, consider how a company's culture will affect you.

Values. By looking at what the company values, you can learn a lot about the company. Does the company want its employees to take the initiative, or does it prefer workers who always ask for orders first? Does the company contribute to the community? The answers may tell you if your personality would fit into the culture.

Goals. What are the company's goals, and how does it meet them? For example, one company may focus on giving customers the lowest price or best value. Another company may emphasize quality and focus on high-end, high-priced custom work. It is useful to know what your company's goals are, because it can help your career. The cost-conscious company will welcome ideas that improve efficiency but probably would reject ideas that increase expenses. The company specializing in custom work wants to improve efficiency too, but probably would not balk as much at ideas that could cost more. If you work for this kind of specialty company, you probably would not want to suggest (for example) replacing a porcelain bathtub with a plastic shower stall.

Climate. How do the workers relate to each other, and how does the company relate to its employees? Are the workers chummy, socializing with each other after work hours and on the weekend? Or do they go their separate ways at the end of the week? Are the company's supervisors promoted from within, or do they tend to get hired from the outside? How strict is management? How flexible are the supervisors? The answers to all these questions will help you determine what the company's climate is like, and what you can do to best fit in.

Tips for succeeding in your new job

When you start a new job, you might feel a bit overwhelmed with so much to learn. Most people know that a new hire needs a period of adjustment. A good starting point is having a good attitude. Here are some ways to help you get off on the right foot:

Take notes when appropriate. When you start a new job, you might feel swamped by all the new information thrown at you: company procedures, the requirements of the job, safety regulations. You cannot really be expected to remember everything—and you probably will not—but you have to absorb as much of it as possible. So make the best effort you can. Take notes when you receive instructions or when procedures are described. It is not always practical to take notes, so don't hesitate to ask that something be repeated if you did not fully understand it. Your supervisor will see that you are serious about doing your job safely and well.

Make a positive first impression, but don't show off. Establishing good relationships with your co-workers is important to your job satisfaction. It requires a bit of a balancing act. On the one hand, you want to show your supervisor that you are paying attention and eager to learn and work hard. On the other hand, you do not want your co-workers to think you are kissing up to the boss. When you ask questions, don't overdo it; ask about what you think you need to know, but don't ask questions just to show off how committed you are. Do not try to flatter the boss, or anyone else for that matter. It will not impress your supervisor, and it might antagonize your co-workers.

Ask questions. Asking questions is the best way to get information. It also shows others that you want to learn and do a good job. Make sure the questions are appropriate and work related. You want information about the job; you don't want to accidently offend a co-worker.

Summary

Keeping your job means acting like a professional. That means being dependable, honest, prepared, organized, and respectful. It also means showing up on time and taking leave only when it's been scheduled or when you're sick. Professional workers always show consideration for their supervisors and co-workers; they know that if they don't show up when they're expected, it affects everyone. As you start your new job, keep in mind that you're a professional and must act accordingly. That way, you'll develop the kind of good reputation that will allow you to progress in your construction career.

Here's a quick quiz that allows you to apply what you've learned in this module. Select the best possible answer, given what you've learned in this module.

1. You've overheard the company owner telling your supervisor to hire only people who have a strong work ethic. The owner means that only people who _____ should be hired.

   a. are technically qualified

   b. believe in giving a day's work for a day's pay

   c. cooperate with their co-workers

   d. never call in sick

2. Most bosses say that _____ are two of the most common problems management has with workers.

   a. alcoholism and depression

   b. absenteeism and lateness

   c. poor skills and language barriers

   d. sloppiness and irresponsibility

3. You are hired to a job in highway construction, and your supervisor says that you must wear an orange vest to be visible to drivers. You've seen other highway crews, though, whose members don't wear those vests. What is your best course of action?

   a. See if other workers are wearing their vests to decide whether you really have to wear one too.

   b. Recognize the reason for the rule and wear your vest.

   c. Ask your supervisor if you have to, because you've seen other highway workers who don't wear vests.

   d. Wear the vest reluctantly, but read the employee manual later and if it doesn't say anything about wearing a vest, tell your supervisor that the next day.

4. Because of a situation beyond your control, you're going to be 15 minutes late for work. You should _____.

   a. start working as soon as you arrive and hope nobody noticed you were late

   b. call your supervisor as soon as you know you'll be late, explain the situation, and apologize

   c. ask a friend to punch in for you, and then work an extra 15 minutes to make up for the time

   d. ask somebody else to call your boss to explain that you're on your way and will be there soon

5. You've made a serious mistake that will cost $1,500 to fix. Because you're friendly with your co-workers, they agree to share the responsibility to take some of the heat off you. You should _____.

   a. accept their offer, because you should cooperate with your co-workers

   b. try to cover up the problem while looking at it as a learning opportunity

   c. thank your co-workers, but take full responsibility for your mistake

   d. say you're sorry to your boss, but defend yourself because the mistake was not entirely your fault

6. Your boss tells you that you have to please many customers in your job. Which of the following would *not* be considered your customers?

    a. Your co-workers

    b. Your supervisor

    c. The owner of the company

    d. The company that supplies building materials for the job you're working on

7. You're working on a kitchen renovation job that must be completed by Friday, since the home-owner is planning a dinner party for the weekend. But it's late on Thursday, and you're sure the project is not going to be finished on time. The project supervisor has left for the day and is not scheduled to be there on Friday. You have the supervisor's home phone number, but you're reluctant to call the boss at home. Your best course of action is to _____.

    a. suggest that the owner cancel the dinner party

    b. contact your supervisor at home Thursday night, explain that there's a significant problem, and propose a solution if you have one

    c. try to contact your supervisor's boss on Friday, if it looks like the project is still off-schedule, and explain that the problem is that your supervisor is not there to oversee things

    d. finish as much of the job as possible before the weekend, even if you have to cut a few corners

8. You're a trainee carpenter who is working hard to become a master carpenter. One day, your boss asks for volunteers to help the masonry crew, which is shorthanded. What is your best course of action?

    a. Volunteer to help out and learn as much as you can about masonry.

    b. Volunteer to help out if your boss treats your work with the masons as overtime.

    c. Explain to your boss that your inexperience in masonry might pose a safety hazard and so you shouldn't do it.

    d. Welcome the opportunity as a mini-vacation from your real job and help out with simple tasks like unloading supplies.

9. Your supervisor has called the work crew together to discuss some rather complex custom work you are scheduled to do. You should _____.

    a. suggest to your boss new techniques you've heard about, which would lower the costs of doing the job

    b. ask lots of questions, to show the boss you care more than anyone else does

    c. take notes and ask about anything you don't understand

    d. stay silent because you're new on the crew, and let the more experienced workers ask any questions or make any comments

10. You come to work after a rough weekend when there was some upheaval in your personal life. You should _____.

    a. complain about your bad luck, and expect your co-workers to empathize

    b. ask your supervisor for a "mental health day" off, to clear your head

    c. stay focused on your job and try not to let personal matters affect your work

    d. go to work but don't do a whole lot, because you've had a tough time recently and your boss and co-workers will understand that you're just not into working today

Individual Activities

Activity 1: Making the Right Impression

Read the following case study, and then answer the questions that follow.

Case Study: New on the Job

You have landed your first construction job with a company that builds warehouses, silos, and other storage facilities for the agriculture industry. You are so excited to start work the next day that you can't sleep. When your alarm buzzes, you hit the snooze button a couple of times, then grab a quick shower, brush your teeth, and comb your hair. A glance at your watch shows you are running only about 10 minutes late.

At the jobsite, your supervisor introduces you to the owner of the company. You shake hands, say how happy you are to have this job, and promise to give 110 percent. You say that this job is a dream come true and talk about how you always played with tools as a kid. The owner smiles and starts to move away, but you keep talking for a few more minutes.

Your supervisor hands you a broom and tells you to sweep the floor. You take the broom, roll your eyes, and sigh in disappointment and disgust: they didn't say anything about sweeping floors when they hired you! Nevertheless, you buck up and do what you've been told. As you sweep, you notice several tools lying on the floor against the wall, so you put them back where they belong. You also notice that the broom handle feels a little loose, so you tighten it. When the trash bin is full, you move it out to the trash collection area.

You've been asked to finish the job in one hour, but you are a fast, neat worker and finish 15 minutes early. Because you missed breakfast, you relax with a candy bar for 15 minutes. Then you report to your supervisor, who introduces you to a co-worker assigned to train you on a new piece of equipment.

Your trainer doesn't seem to want to do this and is a bit grumpy. You feel uneasy, so you don't pay attention during the training. Luckily, it's lunchtime soon, and you leave just before the trainer is finished discussing how to erect scaffolds when building a silo. At lunch, you buy soft drinks for four other new workers to celebrate being hired. You warn them to avoid the grouch who is training you, and you get so involved in talking about it that you lose track of time. Though you are only a few minutes late getting back from lunch, you sense that your trainer is annoyed. After you make several mistakes on the new equipment, the trainer lectures you about not being on time and not paying attention. You apologize for being late and for not listening, and promise to try harder, but it puts you in a foul mood for the rest of the day. After all, you're tired because you didn't get much sleep the night before. The next morning when you wake up to go to work, you're not nearly as enthusiastic.

List at least four things you did right.

1._____

2._____

3._____

4._____

List at least four things you could improve.

1._____

2._____

3._____

4._____

Think about what you could do differently to make your second day better than your first.

Activity 2: Projecting a Good Attitude

What is a good attitude? It may be best to define it by what it is not. The following chart lists behaviors that most people think characterize someone with a bad attitude. If you check Never or Rarely by most of these, that indicates you have a good attitude, and you'll get off to a good start at your new job. If you check Sometimes or Often by most of these, however, that indicates your attitude could be considered negative, and you should work on improving it. As you do this activity, evaluate yourself honestly.

		Never	Rarely	Sometimes	Often
1.	I show up late for work.	○	○	○	○
2.	I complain about my workload.	○	○	○	○
3.	I am sarcastic about the job, my supervisor, or my co-workers.	○	○	○	○
4.	I usually know better than my co-workers.	○	○	○	○
5.	I shift the blame for my mistakes.	○	○	○	○
6.	I generally work harder than everyone else.	○	○	○	○
7.	I don't accept criticism gracefully.	○	○	○	○
8.	I have a short temper.	○	○	○	○
9.	I use obscene language at work.	○	○	○	○
10.	I don't think keeping to schedules is that important.	○	○	○	○
11.	I take it easy on the job at any opportunity.	○	○	○	○
12.	I stretch out my lunch breaks, and clear out early as often as I can.	○	○	○	○
13.	I'd rather laugh at others than at myself.	○	○	○	○
14.	I treat others with less respect if they're a certain race, color, religion, age, or ethnicity.	○	○	○	○
15.	I treat others with less respect due to their gender or if they are disabled.	○	○	○	○
16.	I don't like people telling me what to do.	○	○	○	○
17.	I do my own thing and if my co-workers or boss complain it affects the job, the heck with them.	○	○	○	○
18.	I don't clean up my work area when I finish working.	○	○	○	○
19.	I disregard safety rules if I don't see the point.	○	○	○	○
20.	I say, "That's not my job."	○	○	○	○
21.	I don't like to be told I've made a mistake.	○	○	○	○
22.	I argue a lot just for the sake of it.	○	○	○	○
23.	I talk when I should keep quiet.	○	○	○	○
24.	I get in a sour mood if things don't go my way.	○	○	○	○

Now that you've filled out the checklist, select two or three behaviors that you checked as Sometimes or Often, and come up with an action plan to help you improve that behavior. A sample action plan, for improving the habit of showing up late for work, is provided for you to use as a guide.

Action Plan for Improvement

Sample Problem: *Showing up late for work.*

Sample Action Plan:

1. Go to sleep half an hour earlier and wake up half an hour earlier.

2. Buy a reliable alarm clock.

3. Check the news for traffic or weather conditions that could delay your commute.

4. Ask a co-worker to help you get to work on time.

Problem: _____

Action Plan: _____

_____

_____

_____

_____

Problem: _____

Action Plan: _____

_____

_____

_____

_____

Problem: _____

Action Plan: _____

_____

_____

_____

_____

Activity 3: Surveying the Company Culture

This activity will help you better understand your company's culture: its values, goals, and climate. It may also give you ideas for dealing with situations you're not sure how to handle. On a separate sheet of paper, write down answers to the questions to the Workplace Culture Questionnaire below. Base your answers on the company you work for or, if you don't work for a company yet, on the kind of company you'd like to work for. As you fill out the questionnaire, write down any workplace culture situations you've experienced at your job (or situations you think you might experience) that you're not sure how to handle.

Workplace Culture Questionnaire

1. What are some of the standard behaviors or practices in your workplace? What does the company value?

2. What are your company's goals, and how does it attempt to meet those goals?

3. What is the atmosphere at the workplace like? Easygoing or strict? Are people generally talkative or quiet? Serious or humorous? Stressed-out or relaxed?

4. Name one behavior that is acceptable in your workplace.

5. Name one behavior that is unacceptable in your workplace.

6. How did you learn which behaviors are acceptable at your workplace and which aren't?

7. Do you often hear people saying things like "You can't do this" or "You shouldn't try that" or "Never suggest that here"?

8. Do you often hear people saying things like "That sounds like a good idea" or "What do you think about that?" or "We could do it that way"?

9. Does your company treat everyone equally, regardless of their race, color, gender, religion, national origin, age, or disability? Does your company seem to not hire people based on these characteristics?

Now, break into groups of three or four to discuss the similarities and differences among your team-mates' answers. Discuss the various workplace culture situations you and your teammates listed.

Team Members

1. _____ 2. _____

3. _____ 4. _____

Activity 4: Examining the Effects of a Negative Attitude

We've stressed the importance of having the proper attitude throughout this module. People may sometimes differ on what constitutes a positive attitude, but there are a number of behaviors that most everyone agrees reflect a bad attitude. In this activity, you will examine the effects some of these behaviors—behaviors you should avoid or, if you have them, change. To complete the chart, work in groups of three and assign roles as follows:

Team Members

1. _____

2. _____

3. _____

1. _____ plays the part of a worker with a negative attitude.

2. _____ plays the part of a co-worker.

3. _____ plays the part of a supervisor.

How Workers with Negative Attitudes Affect Themselves, Their Co-Workers, and Supervisors

(The first row has been completed to get you started.)

Problem	Effect on Worker with a Negative Attitude	Effect on Co-Workers	Effect on Supervisor
1. Often argues with co-workers about unimportant matters	• Everyone avoids this person • Has no friends on the job • This person becomes more angry as time goes by	• Hurts the work crew's morale • Makes it hard to communicate, which could lead to mistakes • Makes everyone's job less pleasant	• Wastes time being the referee • Draws attention away from more important problems • Causes problems with higher management
2. Often is late or absent			
3. Uses obscene or insulting language			
4. Takes credit for others' work			
5. Gossips or spreads rumors			
6. Is lazy, does not pull their own weight			
7. Always blames others for mistakes			

Activity 5: Examining the Effects of Absenteeism and Tardiness

In this activity, you will work in groups of four to examine the impacts of absenteeism and tardiness. Imagine that you and your team are assigned to a challenging welding project. The work area is located on a busy downtown street with many people walking past. Your boss has given you 4 hours to get the job done.

All four of you must be at the jobsite on time, for the full 4 hours, to get the job done on time and on budget. Assign the following roles and then answer the discussion questions.

Team Members

1. _____ 2. _____

3. _____ 4. _____

1. _____ plays the part of the worker in charge of getting the work area ready. You must put up safety barriers and keep passersby a safe distance from the welders.

2. _____ and

3. _____ play the parts of the welders.

4. _____ plays the part of the worker responsible for responding first if something goes wrong. You must stand by with a fire extinguisher.

Discussion Questions

1. Suppose one of the crew doesn't show up for work. What is likely to happen? Will the job get done on time? How can the remaining three team members rearrange themselves to get the work done safely and on time?

2. Now suppose two of the workers on the team do not show up. Can the job be done at all? If so, how? If not, why not?

3. When workers don't show up or show up late, how is safety affected?

4. What will happen if one worker leaves early? What if two workers leave early?

Building a Strong Relationship with Your Supervisor

"That's my main flaw: I always think authority figures or my boss is going to think something I do is funny. [And] usually they don't."

– Jimmy Kimmel

Introduction

Everyone has a boss. Even the owner of the company you work for has a boss: the customer. Many bosses have started in the same position you are in and have earned their current positions. As you continue to gain new skills and experience, you may want a job with more responsibility and higher pay. For now, you must focus on your current job and remember that before you can lead, you must learn to follow.

If you work hard and act responsibly on the job, most supervisors will respect your efforts and will sympathize with the challenges you face. At some point, your supervisor has probably done your job or something similar. Because you are just starting out, it's not likely that you'll know something about your job your boss does not. That can work to your advantage; your boss can be your mentor and help you pursue your career. You can always ask your supervisor for advice on how to do a better job and what additional training you might need.

As you pursue your career in construction, you may come across bosses who you think are difficult to get along with, but you should always respect the job they are trying to do. You might find it difficult to put yourself in your supervisor's shoes, but no matter where you work, much of your success will depend on how well you get along with your boss. In this module, you'll read about what a supervisor's job demands, and then you will learn how to build a positive relationship with your boss.

> "By working faithfully 8 hours a day, you may eventually get to be the boss and work 12 hours a day."
>
> – Robert Frost, American poet

Understanding your supervisor's job

Bosses must deal with not only those they supervise—you and your co-workers—but also with their supervisors. They must also work with material suppliers, inspectors, clients, and others. Supervisors are responsible for the big picture as well as all of its individual pieces. Supervisors must coordinate all of the tasks on a project and ensure that employees are working together sensibly, efficiently, and safely. Supervisors must take a broad view of every issue and make decisions based on what is best for the project overall. When you are dealing with your supervisors, keep in mind all these responsibilities they have.

In general, supervisors are responsible for the following:

Safety. Supervisors are in charge of safety at the worksite, ensuring that work procedures are not dangerous, equipment is safe to use, and all workers know and apply safety rules. Safety is the responsibility of every worker, but it is a particularly heavy one for a supervisor. If there is a safety-related incident, the site supervisor is accountable. Both state and federal laws govern workplace safety. Companies that do not follow workplace safety laws risk losing workers to injury or even death and can expect to pay huge fines, in the thousands or even millions of dollars. Construction companies take safety very seriously and demand that their supervisors carefully oversee this important task.

Productivity and quality. Supervisors make sure that quality work is done on time. Site bosses see to it that workers are using the proper tools, equipment, and procedures. Supervisors are always watching to make sure that workers are not wasting time or supplies, and they are always looking for ways to do the job better or more efficiently.

Coordination. Supervisors make sure the project runs smoothly and that work is not delayed. They must schedule workers and tasks so that the different trades do not get in each other's way. Supervisors must also ensure that workers do not have a lot of down time; it is too expensive to have workers idled because supplies have not been delivered.

Cost control. Construction is a profit-oriented industry. To be profitable, a job must come in at or under the budgeted costs. Supervisors make sure that money isn't wasted and that contractors do not over bill and suppliers do not overcharge. Because time is money, supervisors watch to make sure that workers are reporting when scheduled, taking a reasonable amount of time for meals and breaks, and putting in a full day's work.

Leadership. The boss at a construction site is responsible for ensuring that the right materials and tools are available, that the working conditions are safe, and that the workers are doing what they are supposed to be doing. A boss who is a real leader does even more: motivates the workers and makes them feel as if they're important members of the work crew.

You have a much more focused job than your boss does. You work on specific tasks, and you do what your boss tells you to do. By doing your tasks professionally, you can help make your supervisor's job easier. Remember, supervisors have responsibility for all crew members, not just you. Therefore, supervisors who know they can rely on you will come to appreciate you, because they will know they don't have to worry about the job you are doing. Then they can deal with the many other responsibilities that come with their position.

Tips for building a strong relationship with your supervisor

Most supervisors are confident and have strong personalities. They must have these qualities to handle their bosses, workers, suppliers, inspectors, and clients. They must deal confidently with the pressures that come with being in charge.

Not all supervisors belong where they are. Someday, you may have a boss you believe is ineffective. That does not give you the right to refuse to follow your boss's orders; it does give you the opportunity to learn from that boss's mistakes. Even if you don't agree with your bosses, you need to develop good relationships with them.

Here are some suggestions for getting along well with your boss.

Remember that your boss is not the enemy.

- Bosses don't spend most of their day trying to make your life difficult. It is their responsibility, however, to supervise their workers, which means telling them when they do something wrong.

- Don't take it personally if your boss reprimands you, no matter the situation.

- Bosses don't generally single out individual workers, because it's not professional for a supervisor to show partiality for or against any employee.

- Take a moment and think about how you act around your boss. You want your boss to be honest, fair, open-minded, and even-tempered. Do you act that way around your boss?

Respect the boss–worker relationship.

- Your boss is not there to be your friend. There must be some distance between the boss and the worker for their relationship to stay professional.

- Most bosses and workers appreciate that their relationship should be cordial and based on mutual needs.

> "Friendship's got nothing to do with it."
>
> – *Miller's Crossing*

- If you come across your boss (or your boss's supervisor) outside of work, be pleasant but don't act as if the person is the brother or sister you haven't seen in ten years.

Offer solutions, not complaints.

- Your boss has authority over you in the workplace. When your boss tells you to do something, you must do it without complaint.

- If you disagree with your boss because you think you have a better way of doing what you've been told, make a suggestion. Instead of griping, "I don't want to do this your way," offer an alternative: "I figured out a way to do this that I think will save time, boss. Can I show you?"

- There might be times when you have a justifiable reason for not following instructions. If this is the case, state those reasons calmly and respectfully.

Be flexible.

- Just because you're learning to be a specialist in one trade does not mean you should refuse to help in others. There's always something new to learn.

- If your boss asks you to help at another task or work at a different site, treat the request as a learning opportunity.

- It never hurts to add to your skill set; you'll command more pay that way, and you'll develop the scope of skills that may allow you to be a boss one day.

Communicate wisely.

- There are times when you should talk to your boss, and times you shouldn't.

> "Many attempts to communicate are nullified by saying too much."
>
> – Robert Greenleaf, management expert and author

- If you want to speak with your boss, especially about something involved, pick your spots carefully. If your boss has had a stressful day, it might not be the best time to ask for time off or a raise.

- Unless the matter is urgent, it's not a good idea to interrupt your boss in the presence of your boss's supervisor or the company's customers.

- Some problems will not wait. If there is a pressing matter you must bring up with your boss in the presence of higher management or the company's clients, interrupt as politely as you can and ask if the boss can step aside for a few moments to help deal with a serious issue.

Follow the chain of authority.

- Another part of acting tactfully with bosses is not going over their heads when you have a problem. Bosses often resent that type of behavior.

- If you have problems with co-workers, you should try to work them out on your own. If you can't, give your boss a fair chance to resolve any problem.

- If it's a major problem, go over your supervisor's head only as a last resort and only if you have a very good reason for doing so. First, check with your company's human resources department, if they have one, before you make a direct appeal to your boss's supervisor.

Deal with your mistakes.

- Everyone makes mistakes; the trick is to avoid them as much as possible.

- When you do make a mistake, admit it, take responsibility, and work to limit any damage. Put your effort into correcting the mistake and less into apologizing for it.

- Do not try to cover up a mistake. Covering up a mistake could create a safety problem for you, your co-workers, or the people who will use what you're helping to build.

> "Experience is simply the name we give our mistakes."
>
> – Oscar Wilde, author

- If you admit to the mistake and take responsibility for it, your boss may be angry initially but later will probably respect your honesty. If you try to cover up the mistake and your boss finds out, however, you'll be in double trouble: over the mistake, and over the attempt to cover it up.

Respect your boss around co-workers.

- It may be tempting to complain about your boss to your co-workers, especially after a hard day. Avoid the temptation. Even if the complaint is in confidence, chances are it will be repeated and eventually get back to your boss.

- If you have a question about your boss's methods, speak to the boss in private. You don't want to undermine their authority in front of the rest of the crew.

- Don't spread rumors about your boss, not only because the boss may find out that you are involved; you need to have enough respect for yourself and others to not repeat rumors.

Empathize, empathize, empathize.

- Empathy is one of the most important skills you can have.

- If your boss is reprimanding you, it may be hard for you to put yourself in your supervisor's shoes. It's important to remember that while you see your job from your point of view, your boss looks at it in terms of how it affects the overall project.

- If you are late, waste materials, or do a poor job, you're having a negative impact on that project, and it is your supervisor's duty to reprimand you and get you back on track.

Be proactive.

- Knowledge may be power, but it's enthusiasm that turns on the switch.

- Showing dedication and eagerness are two outstanding ways to start your relationship with your supervisor on the right track.

- Enthusiasm should be balanced by respect. If you find an error in a co-worker's job, for instance, you might think the best thing to do is go ahead and fix it without talking to your supervisor. A better course of action would be to bring the matter to your boss's attention.

Nonverbal Communication

You may think you're saying all the right things to your boss, but your body language may be sending an entirely different message. "Right away, boss" doesn't sound too convincing if you're shaking your head or rolling your eyes while you say it. So, when you talk to your boss, be aware of your body language. Stand up straight and don't slouch, avoid crossing your arms and clenching your fists, look your supervisor directly in the eye, and don't frown. If you mean what you say, it'll show in your body language.

Workplace politics

When you spend eight hours or more a day at work, it's unavoidable that you'll want to influence things at the jobsite in your favor. But it's easy to go too far. *Workplace politics* describes the behavior of employees who try to manipulate events in the workplace by making trouble: for instance, by talking about people behind their backs. Workplace politics can be fairly complicated and often frustrating, but there are two simple and effective ways to avoid getting involved: Don't do or say anything that makes your boss look bad with upper management, and behave professionally at all times.

Here are some aspects of workplace politics and suggestions to help you avoid them.

Deal with troublemakers with deeds, not words. A worker with a bad attitude is like a virus; one person can infect everyone else. Most people want to avoid troublemakers, but that's not always possible. Direct confrontation can lower you to the troublemaker's level of childish behavior and often makes the situation worse.

- Encourage a positive environment by staying upbeat, so that your good attitude will rub off on those co-workers who might be susceptible to those with a negative attitude.

- Do not become a sounding board when someone starts bad-mouthing the boss or anyone else. Negative people need an audience. Don't provide one.

- Lead by example. Point out that there is work to do and then start doing it. Others will follow.

- Do not get pulled into causing trouble yourself. People with bad attitudes often want to recruit others into their schemes. Don't let that happen to you.

Don't be a pest. You are not the only worker who must deal with the pressures that come with work and your personal life. Others will be sympathetic to your problems at first, but if you become known as a whiner, they will soon avoid you.

- Problems are a part of everyone's life. You may think your problems are worse than everyone else's, but other people likely feel the same way about theirs. Keep yours in perspective and, to the extent you can, to yourself.

- Instead of griping, take positive action. If you don't like the tasks you're being given now, get training to do the ones you want. Ask your supervisor for guidance. If your company has a human resources department, make use of it. It's there to help a company's employees improve their skills and advance in their careers.

- Be patient. You may not like a lot about your job now, but think of it as a stepping stone to better things. When you show your boss that you can handle any task well, you will find it easier to get raises and promotions.

Don't take advantage of people. If you do not respect your supervisors or if you try to take advantage of them, they will soon lose respect for you.

- Be considerate. Do not take advantage of your supervisor's good nature by slacking off on the job.

- Be thoughtful. Keep your supervisor informed regarding your availability. Do not leave a supervisor hanging and wondering where you are.

- Be professional. Avoid goofing off or playing practical jokes on others. What may seem like harmless fun to you could cause a problem for your supervisor.

Avoid the rumor mill. Say you're on break with a few of your co-workers, and they start spreading stories about the boss, gossiping about the company's president, or taking cheap shots at the new hires. That's when you know the rumor mill has started, and that's not a place you want to hang around.

- Do not pass on what you've heard to someone else. Even if you repeat a rumor disapprovingly, you're still repeating it, which means you're still on the rumor mill circuit.

- Respond to insincerity with sincerity and good intentions. You'll develop a reputation as a stand-up person, which is the kind of worker supervisors like to have around.

Summary

Getting along with your boss is vital to your success, and it makes things a lot easier for you as well. You can build a positive relationship with your supervisors by understanding their jobs and responsibilities, respecting them, offering solutions and not complaints, and accepting responsibility when you make mistakes. Remember, your boss is not there to be your friend, but your boss is also not your enemy. By keeping all this in mind, you'll have good relationships with your supervisors, who'll think highly of you.

On-the-Job Quiz

Here's a quick quiz that allows you to apply what you've learned in this module. Select the best possible answer, given what you've learned.

1. Your supervisor, who you respect and admire, is moving to another state to take a new job. A few weeks later you meet your new supervisor, who is coming from a rival construction company. You should assume that _____.

   a. your new supervisor will know next to nothing about the way your company works and you will have to babysit

   b. the new boss is bound to be a hard case given that your old boss was so good

   c. your new supervisor is skilled and capable, but probably a bit nervous about starting with a new crew and company, so you should be as helpful as possible

   d. you'll have to butter up the new boss so that your skills will be noticed and you'll get that promotion you earned under the old boss

2. You've just started a new job, and you like your new, easygoing supervisor. After a couple of weeks of working hard, you start to slack off. You should assume that your supervisor _____.

   a. will sooner or later casually mention the workplace rules and give you a friendly reminder to follow them

   b. has many responsibilities and so probably won't notice if you stretch out your lunch breaks or leave early now and then

   c. will tell you in clear language to straighten up and follow the workplace rules

   d. is cool and probably feels that as long as the job gets done eventually, a little slacking off is OK

3. It's a quiet Friday afternoon. Everyone is in a pretty good mood because on Wednesday your crew finished a huge job on time and on budget. You have a week of vacation leave, and you want to use it in a couple of weeks to take a trip. Is now a good time to ask your boss for time off?

   a. No, because you've heard a rumor that your boss doesn't like to be asked for time off, and so you should ask someone above your boss.

   b. No, because you should make these types of requests on Monday when the human resources office can handle it.

   c. Yes, because you've earned a vacation, and it's your responsibility to let your boss know.

   d. Yes, because your boss can more easily spare you some time now that the pressure of the big job is over.

4. You've been on the job a couple of weeks, and your co-workers seem to like to bad mouth the supervisor a lot. You want to get along with your co-workers, but you also want to get ahead. You should _____.

   a. listen carefully to what your co-workers are saying, then repeat it to a materials distributor you talk to later in the week to verify if there's any truth to what your co-workers were saying about the boss

   b. avoid bad-mouthing the boss until you've got a good personal reason to do so

   c. avoid bad-mouthing the boss and treat your supervisor with respect

   d. be proactive and tell your boss's supervisor about what your co-workers are saying

5. You hear your boss reprimanding a co-worker who left tools out in the rain overnight. You look at the situation from your supervisor's point of view and decide that _____.

   a. strong leadership means chewing someone out every now and then to let everyone know who's boss

   b. your boss should have privately warned the worker and not humiliated the worker in public

   c. your co-worker didn't know it was going to rain, it was an accident, so your boss must really have it in for that co-worker

   d. tools are expensive, and they shouldn't be left out overnight, whether it rains or not, so your boss was justified in reprimanding your co-worker

6. You're reading a blueprint, and it's not making any sense to you. There seems to be a serious mistake in it, but because you're not that experienced, you're not sure. You should _____.

   a. assume that the person who drafted the blueprint knows more than you do, and follow it exactly

   b. ask your supervisor to look at the blueprint and advise you on what to do

   c. show initiative, ignore the blueprint, and complete the work according to your own judgment

   d. tell your supervisor someone made a mistake on the blueprint and that you can be relied on to set these kind of things right in the future

7. You're feeling pretty annoyed. Your co-worker is supposed to relieve you at 4 p.m. Yesterday, you were not able to leave until 4:03, and today it was 4:04 before you could punch out. You're not getting paid overtime, and you resent having to wait. You should _____.

   a. promptly complain to your supervisor by noting that the tardiness is screwing up the boss's careful work plan

   b. be patient for a day or so and if the problem continues, explain to your co-worker that you need to leave at 4 p.m. each day

   c. complain to your boss's supervisor, choosing your words carefully

   d. tell your boss that you are willing to wait till your co-worker shows up, but that you expect to be paid overtime for any time worked after 4 p.m.

8. You're at the grocery store with your family and you come across your boss's supervisor, who is also shopping there. You should introduce your family and _____.

   a. mention how much you enjoy your job, and then be on your way

   b. offer to help the supervisor carry the groceries to the car

   c. use this as an opportunity to make yourself known to upper management by talking shop and suggesting some ways to control costs at the jobsite

   d. mutter hello and then leave as soon as possible because you don't want a co-worker or your own boss to see you schmoozing with upper management, if they happen to be in the store

9. You're talking to your boss about the trouble you're having getting the permits needed to start a job. However, your supervisor keeps glancing out the door and tapping a pencil on a desk piled with papers. Based on the nonverbal cues you're receiving, you should _____.

   a. ask your boss if there's something the matter

   b. offer to come back later to talk about the problem

   c. tell your boss that you only need a few minutes and to please listen because what you're saying is important

   d. close the door so your boss will stop looking at whatever the distraction is

10. Your first week on the job, a more experienced worker tells you to go get a skyhook and not to come back until you find one, asking everyone on the site if need be. Several hours later, you realize that there's no such thing as a skyhook and that the senior workers were playing a joke. Then your boss reprimands you for not staying focused on your job. You should _____.

    a. apologize to your boss but say it wasn't really your fault because your co-workers were playing a joke on you

    b. start telling others about what irresponsible goofballs those senior workers are and that they ought to know better

    c. admit to your boss that you were distracted by your co-workers and that you'll stay more focused in the future, and then resolve not to be pulled into a prank like that again

    d. complain to the company's human resources department about the senior workers who pulled the prank

Individual Activities

Activity 1: Managing Difficult Situations with Your Supervisor

In this module, we've emphasized the importance of following orders without complaint. There will still be times, however, when simply following orders is not enough, and you'll have to manage your boss—even as your boss is managing you—to avoid conflicts or resolve them if they do come up.

What follows are several resolution techniques you can use to present your case fairly and politely to your boss. Match each situation in the left column with the best conflict-resolution technique in the right column. This can also be a group activity, in that you can get together with your fellow students and compare techniques after you've completed the exercise. Once you've matched a resolution technique to a situation, discuss which bullets for that resolution technique help to address which bulleted aspects of the corresponding situation.

Situation	Resolution Technique
Situation One: Work versus family. • The company has a pressing deadline, and your boss, who's usually a good, reasonable person but in this case is stressed out, wants you to put in 20 hours of overtime during the weekend. • That means you'd miss a long-scheduled family outing on Saturday afternoon; your beloved and ailing grandmother, whom you haven't seen in over a decade, will be there. • You realize it may be your last chance to see her, so you don't want to miss the outing; on the other hand, you do want to help your boss and keep your good relationship going. • You offer to work some overtime during the weekend but not on Saturday afternoon, and work whatever other hours are necessary to help complete the project on schedule.	a. Reasoning: • Use simple facts and data to support your case, while acknowledging it might just be a lack of understanding on your part. • Don't get emotional or lose your temper. Be calm and reasonable to best present your side of the story to get the information you need. • Realize that not everyone gets everything right the first time, not even a boss, and that you should always show your supervisor respect regardless of the situation. This fits best with Situation _____.
Situation Two: "I've got a better idea." • Your supervisor has strong opinions about the specific way a wiring job should be done. • You're an apprentice electrician who's been praised on how quickly you pick things up. Your boss, meanwhile, has a reputation about not knowing much about electrical wiring. • You start talking about the wiring job, and you come to realize your supervisor knows quite a bit more about wiring than you've heard. • So you say, "Boss, say I had a way to get this job done more easily and less expensively; would you be willing to give it a go if I could convince you of its benefits?"	b. Bargaining: • Understand that your supervisor has responsibilities that have to be met. • Understand that what you consider to be important may conflict with what the boss considers to be important; that your personal priorities may conflict with what your boss considers to be your professional responsibilities. • Explain what your priorities are while respecting that your boss has priorities as well. • Trade one thing for another; compromise is often a matter of doing favors for each other; when two people exchange favors, both win. This fits best with Situation _____.

Situation	Resolution Technique
Situation Three: "I don't understand" • Your boss has given you a project with instructions that seem incomplete. • You realize that your boss is busy and hasn't had time to really think the instructions through and may have left something out. • When you ask for clarification, your boss says, "I thought I explained all that to you already." • You calmly and patiently ask specific questions so that you can get all the information you need to do the job correctly.	c. Courtesy: • Accept criticism gracefully and don't take it personally. • Recognize that your boss has skills and experiences you do not. • Compliment your supervisor's techniques and reiterate your willingness to learn. • Remember that your boss has probably been criticized by his or her boss at one time or another and has learned from those mistakes, just as you are learning from yours now. This fits best with Situation _____.
Situation Four: The group wants one thing, the boss another. • You and your team members would like the boss to order a backhoe because, with all the excavations that have to be done at the site, so much time is spent moving earth that there's little time for new building. • The equipment is expensive and your boss will probably not want to order it. For one thing, the project is already a little over budget; for another, your boss has a reputation for being tight with a budget. • The team makes a list of ways that the equipment would help productivity, while showing how the costs upfront would be saved in the long run. • You make the case to the boss respectfully and courteously, while stressing how everyone wants to get the job done and wants to help the boss as much as possible.	d. Experience and expertise: • If you have a lot of experience or expertise in a certain area, you'll be able to explain your solution effectively. Just do so professionally and courteously. • Recognize that even though you've come a long way, you're still just starting out relatively speaking, and your boss probably has more experience than you have. • Be confident of your own abilities, however, and if the answer is "no," don't take it personally. This fits best with Situation _____.
Situation Five: If you've screwed up. • Your boss is critical of your work, which has set back the project schedule and incurred additional costs. • Instead of making excuses or getting upset, you ask how you can improve your performance and then work to improve it. • You express admiration for your boss's skills and that you just want to do the work as you've been told. • You realize the boss has learned from experience. • You also realize your boss doesn't hold a grudge against you, but that part of a supervisor's job is to tell workers when they've done something wrong.	e. Team backup: • Get your co-workers to help you prepare to present your team's case to the boss. • Define the problem and explain why your team's solution would work. • In presenting your case, be as open-minded with your supervisor as you hope your boss will be when considering the suggestion. Recognize what limitations your boss is operating under when it comes to making new purchases. • Never gang up on the supervisor; instead, present the proposed solution as a group suggestion to the group leader, the boss. This fits best with Situation _____.

Activity 2: Responding to Your Boss

In this activity, you will figure out the best way to respond to some things your boss says to you. Write down what you should say and not what you want to say. Remember that on the job, you might not always take a moment to think through what you should say. If your boss says something that's upsetting, you may be tempted to say the first thing that comes into your head, so remember this exercise. Think before you speak. If you're on the job and your boss says something that makes you angry or upset, try counting to five or taking a deep breath before responding. Keeping calm and collected is the sign of a professional.

1. **Your boss:** "How many times do I have to tell you to clean up your work area before you leave for the day? Are you deaf or something?"

   You: _____

2. **Your boss:** "Hey, while you're out, get me a sandwich and a cup of coffee."

   You: _____

3. **Your boss:** "I need you to ride out to the supplier to pick up a bunch of things we need. You may not get back until after quitting time, but it needs to be done, and you're the only one available to do it."

   You: _____

4. **Your boss:** "This is the fifth day in a row you've been late. We don't need slackers around here, so go home and over the weekend, think about whether you still want this job. Then we'll talk Monday. For now, though, get out."

   You: _____

5. **Your boss:** "Listen up, people, the owner of the company is going to show up later today, and given how things have been going, it ain't going to be pleasant. So you all need to be on your best behavior, especially you." (The boss points at you.) "You ticked off the customer the other day, so you're on notice."

   You: _____

6. **Your boss:** "I'm always amazed at the quality of your work. It's consistent, all right: consistently poor."

   You: _____

7. **Your boss:** "That's not a bad idea you've got. But we don't have time to implement it now, for one thing. Also, you do realize how expensive it would be to do it that way, don't you? In the future, we may be able to do it your way. Meanwhile, think about how costly it would be, and come up with some ideas on how to do it for a bit less."

   You: _____

8. **Your boss:** "You've got tickets to the game this weekend? Well, that's a pretty poor excuse to refuse overtime and get ahead in your career, especially when you know this job has to be done by Monday. Now I can't make you come in, but I'll remember it if you don't. So keep that in mind when you decide whether or not to come in on Saturday."

    You: _____

9. **Your boss:** "Look, I know some of the others take longer breaks than you do and that they don't pull their fair share around here. I'm working on changing that but in the meantime, I need you to stay late this week. You're developing a good reputation as being reliable and hard-working, and I want you to keep that reputation."

    You: _____

10. **Your boss:** "There's a right way and a wrong way of doing things. You may not think this is the right way, but I got 25 years of experience in this business, so I'd imagine I know better than you. Come back to me after you've got 25 years of experience, and I'll listen. In the meantime, get to it!"

    You: _____

Activity 3: Name That Attitude

This activity requires four people: one to play the role of the supervisor, the other three to play the roles of workers on a construction site. Read the script aloud, and then discuss the kinds of attitudes the four workers hold toward their boss and what kinds of opinions the boss has toward the workers in return.

Team Members

Supervisor: _____

Workers: _____ Francis

_____ R.P.

_____ Dee

Supervisor:	OK, gang, today's when we start putting up the platform frame…
Dee:	Excuse me, boss, I need next Monday off.
Supervisor:	This isn't the time, ask me later…
Dee:	But this is the only time I can get ahold of you.
Francis:	Dee, shut up. The boss is trying to talk.
Supervisor:	Francis, I can handle this…
Francis:	Well, don't get all upset about it! Gee, I'm just trying to help.
Supervisor:	All right, everyone, cut the nonsense! Now we're going to start by…
R.P.:	What about this lumber? Sorry to interrupt, boss, but the lumber seems kind of flimsy.
Supervisor:	This framing lumber has been properly grade-marked for use under the building codes this project is operating within. So, let me worry about that kind of thing, and you just do your job.
Francis:	Yeah, that's really using your head, R.P.
R.P.:	Don't start with me, Francis.
Supervisor:	Knock it off. Let's get back to the framing.
Dee:	What about my time off? When I ran into Ms. Davis this weekend and asked her for the time off…
Supervisor:	You asked my boss before you asked me?
Dee:	Well, yeah, seeing that you're always so busy and hard to get hold of…
Supervisor:	Don't do that again!
Francis:	Yeah, Dee, can't you see, that embarrasses the boss.
Supervisor:	Francis, I don't need you to do my job for me.
Dee:	Francis worries about everyone's job, especially yours, boss.
Francis:	[*aside to Dee*] I told you that in secret!
Dee:	That's what you think, but when you start backstabbing me…

Supervisor: That's enough. Dee, we'll talk about Monday later. Now, Francis... since you seem to be so concerned about how I do my job, have you got any bright ideas on how to do things better?

Francis: Well, I got a few good ideas on how to go about this framing. We could have a drink after work, I'll buy, and I can tell you about them, I've been reading this book on it.

Supervisor: No, Francis, I know how to do the framing. That's what I was going to tell you all about before all this nonsense began.

Discussion Questions

1. What bad attitudes do Francis, R.P., and Dee have? What good attitudes do they have?

2. How could Francis, R.P., and Dee improve their relationships with the supervisor?

3. What attributes of the supervisor are demonstrated in this scene?

4. What do you suppose the supervisor thinks of Francis? Of R.P.? Of Dee?

TOOLS FOR SUCCESS: CRITICAL SKILLS FOR THE CONSTRUCTION INDUSTRY

Teamwork:
Getting Along
with Your Co-Workers

"I got a simple rule about everybody. If you don't treat me right—shame on you!"

– Louis Armstrong,
American jazz musician

Can you imagine a construction project that could be handled by only one person? Workers with many different skills—designers, plumbers, electricians, carpenters, welders, sheet metal mechanics, and general workers, to name just a few—must cooperate to get a project done. Plumbers can't bury water lines if the backhoe operators don't show up. Electricians can't pull a wire unless they have someone to feed wire into the conduit. Construction projects are complex and construction sites can be dangerous, so teamwork is essential to getting all the component jobs of the project done safely, so that no one gets hurt and the project stays on schedule.

Teamwork is much more than simply working alongside other people. In successful teams, all the workers actively contribute. Workers must respect and help one another because they depend so much on each other. A self-centered egotist or someone who cannot work with others has no place on a construction team; the work the team does is too important and potentially too dangerous to be able to afford someone who is selfish or disrespectful. Even if you're shy or a bit of a loner, you can still get along with others on your team and work well with them.

Think of a successful professional basketball squad: a group of athletes who have always stood out and could easily think of themselves as being above it all. But in the most successful professional teams, all the players, no matter how individually talented, know they must work together to win games. At the construction site, you "win a championship" by getting the project done safely, on time, within budget, and to the satisfaction of the client. That requires everyone to participate and work together.

> "Talent wins games, but teamwork… wins championships."
>
> – Michael Jordan

In this module, we outline several ways to develop good working relations with your co-workers so that you can form an effective project team. These include looking for solutions instead of causing problems, helping your teammates in whatever way you can, and supporting and respecting your co-workers. It's particularly important that you treat everyone with respect regardless of age, race, color, gender, sexual orientation, religion, ability, and ethnic background. It's also important to remember that people from different cultures may react differently to certain situations, and that part of being a good worker is being conscious of and sensitive to cultural and gender differences.

Career Insight

A lot of construction projects involve homebuilding; it's likely you'll be a part of several such projects during your career. Teamwork essentially means cooperation by a group of people as they pursue a common goal. You and your co-workers are one kind of team, which builds a house according to the specifications, schedules, and budgets agreed on at the beginning of the project, and which are relayed to you by your supervisors. Another, larger team—one that you are also a part of—is composed of the parties who agreed on those conditions. These parties can be roughly described as the company responsible for building the house, the individuals or company that designed the house, and the customer who hired the builders and designers.

- The homebuilding company employs your boss, your co-workers, and you. The company wants to make a profit by completing the house as quickly as possible, but it also wants to do top-quality work so that the customer is satisfied.

- The customer can be the person who will occupy the house once it is built, or a company that will sell the house to a homeowner (sometimes the builder and the seller are the same company). The customer needs the house built, or to be ready to be sold, by a certain date, but customers also need to work with the builders and designers so that the house is built the way the customers want.

- The designers of the house may work for the homebuilding company or may have been separately hired by the customer. The designers want to please the customer, and they also want to ensure their work, schedules, and expectations are aligned with those of the builder.

Once these three parties have agreed on the project's outlines—what the size and features of the house will be, how long it will take to build it, and how much it will cost—the building company uses these guidelines as it assigns workers and supervisors to the project. Your boss, in turn, manages your work team in order to meet the schedules and do the work safely and well. On some projects, you and your teammates may be the only crew on a project; on other, larger projects, you might be one of several different work crews participating.

Becoming a team player

As you begin your construction career, you'll be part of a team almost from the start. Being a team player is a skill you'll need right away. How can you become a valuable team member? By respecting, helping, and supporting your co-workers, by offering solutions and not causing problems, and by always keeping the ultimate goal of the project in mind.

Team members respect one another. You should always respect your co-workers regardless of their skill level, education, or background. Disrespecting others is insulting, insensitive, and inexcusable. The best guarantee against acting disrespectfully is to always think in terms of showing others consideration. How can you do this on a regular basis?

- Let others state their ideas or views without interruption. Your co-workers have unique ideas and thoughts just like you do; remember that the best solutions tend to come from a diversity of ideas.

- Don't do anything to make co-workers look foolish in front of others. Don't belittle them or their opinions.

- Be pleasant. Don't be quick to anger or judgment.

- Don't speak negatively to or about others because of their race, color, sex, religion, national origin, age, sexual orientation, or disability.
- Remember the golden rule: Treat others as you would like them to treat you. How would you feel if someone treated you with disrespect? How would you feel if you always had to stay after quitting time because a co-worker always left early?

Team members help one another. "That's not my job" is not something employers or supervisors want to hear. It's also not something co-workers like to hear. You should always be willing to help a co-worker if asked; as a member of a team, helping out is part of your job. Although every member of a construction team has a specific job to do, a good team member can find the time to lend a helping hand and will be glad to do it. Remember, one good turn deserves another. Someday you'll need help from your teammates. Would you want them turning their backs on you?

Team members don't cause problems—they look for solutions.
Sometimes it's easy to let disagreements rule the day. People lock horns over an issue and it becomes a problem. Dedicated team members don't let their differences drive them apart. Rather, they work together to find solutions. Say, for instance, you've got a new team member for whom English is a second language. At first, it may be a bit difficult to understand your new co-worker, but through mutual effort you'll find a way to communicate and to get the job done. Through working together, you'll get a good feeling from overcoming this communications challenge, and you'll probably learn an interesting thing or two in the bargain.

> "Once a word has been allowed to escape, it cannot be recalled."
>
> – Horace, Roman lyric poet, first century B.C.

Team members support one another. Team members often rely on one another for emotional support. There are times when one of your co-workers may be having a rough day on the job or some personal problems at home. A considerate team player makes allowances for a co-worker at such times. That might mean giving your co-worker more space than usual, or leaving the person alone, or offering an encouraging word or two. This doesn't mean you should feel so sorry for a co-worker that you end up doing all that person's work; if you find that's the case, your co-worker is taking advantage of you. A little emotional support is fine, but a good teammate doesn't do someone else's job for them.

Team members are committed to a project. Good team members keep their eyes on the prize—building that addition to the hospital, widening that highway, refurbishing that apartment complex—even if the tasks they've been assigned are modest, such as moving a load of lumber or sweeping out the construction trailer. Whatever the task, all good team members know their actions should be geared toward meeting the goals of the project. When faced with questions or problems on the job, as you decide on your best course of action ask yourself, "Will this help accomplish our goal?" If the decision benefits only you and not the goals of the project, then it might not be the best decision to make.

Tips for getting along with your co-workers

There aren't many people who are cut out to be solitary in their lives or work. Most people need to interact with others. To have a friend, you must be a friend. A similar idea is crucial to having good relations with your fellow employees: to have good co-workers, you must be a good co-worker. How can you be a good co-worker and avoid being a bad one? Check out these tips of what to do and what not to do.

Do's

Realize that you are not the only one who works hard. When you are busy on the job, it's tempting to forget that others are working as hard as you are. Remember that your co-workers are as dedicated as you are and that you all play important roles in completing the project.

When appropriate, praise your co-workers. Everyone likes to be complimented on a job well done. Tell your co-workers when you're impressed by their work. A compliment costs nothing, and it spreads goodwill. Think of how good you feel when someone compliments you.

Share the credit and take responsibility for mistakes. Nobody likes the person who hogs all the credit when something's turned out well. Share the credit you receive, and you'll gain your co-workers' respect. The flip side of spreading the credit around is taking responsibility for your mistakes. Don't expect your co-workers to cover for you if you make a mistake. Taking responsibility for your mistakes strengthens your relationship with your co-workers.

Recognize the contributions of others. Understand that different people bring different talents and skills to the job. There will always be those who are more skilled than you are and those who are less skilled. Learn from those who are more skilled, and be patient with those who aren't at your level of ability.

Meet your deadlines. Always remember that there are workers who cannot start their tasks until you have finished yours. Complete your work on time. Meeting your deadlines is more than just showing consideration. Construction schedules, which are often tight, are tied to budgets, and work delays can cost your company thousands of dollars. If you don't meet your deadline, your co-workers may not be able to meet theirs, and the whole project gets further and further behind schedule.

Realize everyone can feel stress. People get stressed out over their work regularly. Don't let work stress affect how you treat your co-workers. Don't snap at them or get indignant with them. Remember, your co-workers might be under as much stress, or even more, than you are. Instead of taking your aggravations out on your co-workers, think of them as people you can relate to.

Make a good first impression. Be friendly and act professionally from the start, whether at the start of a new job or the start of a new workday. Starting a new job can be stressful, and the best way to reduce that stress is to make a positive first impression.

A Special Tip: Keep Your Problems at Home!

You can't stop personal problems from affecting how you feel. However, you must try to keep them from affecting your co-workers. If you've had a big argument at home, don't bring negative feelings to work. Your co-workers will not appreciate it if you try to make them your scapegoats. If you feel unable to leave your personal problems behind for the day, ask a co-worker, one you trust, if you can discuss your problems during mealtimes or a break. Remember, you are partly responsible for the mood, safety, and success of your team. When you learn to manage your personal problems, you also boost your performance and your reputation as a valued member of the team.

Don'ts

Don't boast or act like a know-it-all. This behavior will annoy your co-workers. The braggart, the know-it-all, and the loudmouth are usually among the least popular people at the worksite. Who wants to listen to people bragging about how wonderful they are, how many dates they get, how well they can fight, or cook, or dance, or grow petunias? Would you want such people on your team?

Don't gossip about your co-workers. People are entitled to their privacy, and gossip violates that privacy. The rumors spread by gossip are usually false and often malicious. Gossiping can hurt feelings and damage relationships. Gossiping won't help you establish yourself with your co-workers; instead, it makes you look unprofessional and petty. How would you feel if someone was saying nasty things about you behind your back?

Don't put other people down. Don't pick on co-workers. Don't insult their appearance, their intelligence, or anything else. Putting others down doesn't just hurt their feelings, it hurts the whole team and poisons the environment. Before you start teasing or mocking someone, take a hard look at yourself. Is there anything about you that someone could make fun of? If you laugh at others a lot, don't expect much sympathy if they start laughing back at you.

Don't shirk your responsibilities. Work has to get done, and if you are irresponsible and don't do your share, your co-workers will have to pick up your slack. They'll resent it, and they'll be justified. You won't last long on the team if you don't do your part.

Caution

There may be the rare occasion when you have to report the conduct of a co-worker. The co-worker might be doing something illegal or something that is endangering your safety or that of your co-workers. You should report this person, however, only if you are reasonably sure that your supervisor is not aware of the behavior.

Assisting others

At some point in your career, you'll be asked to teach others what you know. Maybe you will have to explain a safety procedure or demonstrate how to operate a piece of equipment.

Training others on the job is an excellent opportunity to build good relationships. Most people appreciate and respect a good teacher—especially if that teacher is well-informed, patient, and understanding. Being a good teacher requires several different skills.

Be aware that everyone learns differently. There are at least four different ways that people learn: through pictures, by reading, by listening to an explanation, or through hands-on experience. Although we use all four styles as we learn, many of us prefer one method to the others; some people learn things more quickly by reading about them, others learn more readily by listening to an explanation. Which one works best for you?

- **Visual.** You like looking at a diagram or a drawing to understand how something works. You often use pictures to help explain a concept to someone. You're probably good with blueprints.

- **Descriptive.** You prefer words to pictures. Written, step-by-step instructions are best for you, so you can refer back to them as you proceed. You're probably good with owners' manuals.

- **Auditory.** You prefer to hear an explanation and then write it down in your own words. You probably have a good memory for spoken instructions.

- **Tactile.** You are a hands-on learner. You learn best when you actually get the chance to operate the piece of equipment. You might like to work with models.

Your preferred style of teaching might not match your co-worker's preferred style of learning. Be flexible enough to change your teaching style to better fit how your co-worker learns best. For instance, if you don't seem to be getting the point across by describing it, try drawing a picture instead. Before you start training the co-worker, ask if he or she has a preference.

Feel honored. If your boss asks you to train someone, you should feel proud. It's an acknowledgment that your supervisor has confidence in your skills and abilities. The trust your supervisor is placing in you to train your co-worker should inspire you to be the best teacher you can be.

Be patient. It's hard to stay patient when you're teaching someone who doesn't quite get it. If the trainee continues to have difficulty, consider whether you've been speaking too quickly. Perhaps you're expecting too much too soon from the trainee. Sometimes you may have to repeat something more than once, or show your co-worker how to do something several times. The person you're tutoring will appreciate your patience. If you start to feel frustrated, take a short break.

Teach by example. The best way to teach something is often by demonstrating the steps for completing a task, and then asking the trainee to repeat those steps or to perform them for another co-worker while you observe. This is a form of on-the-job training, which is a method widely used in construction and other professions. There are several advantages to teaching by example:

- You can see if the trainee does the task as it's supposed to be done, not by accident without following the correct procedures.

- The trainee gets to practice doing the task and experiences the actual steps that make up that task.

- Both you and the trainee develop a better understanding of each other, which will allow you to work together better both now and later as teammates on a project.

Offer encouragement. Sometimes people don't have much confidence in their abilities, especially when they're trying to do something they've never done before. Encouragement is a big help, because it gives people confidence. Praise, like telling your co-worker "good job," is always welcome.

Don't be afraid to give constructive advice. To do a proper job of training others, you have to point out when they make mistakes. You can do it harshly, or you can do it gently. People take constructive advice to heart, but they resent advice delivered with scorn or contempt. If the trainee makes a mistake and you have to point it out, do it respectfully. Calmly tell your trainee what the mistake was, and give the trainee a chance to do it again.

Summary

Respecting your teammates, lending them a hand when asked, offering your co-workers praise and encouragement, showing your team members consideration, looking for solutions instead of causing problems, sharing the credit, and taking responsibility—these are the attributes of a valuable and popular co-worker, the kind of person others want on their team. Making fun of people, gossiping, bragging, shirking responsibility, refusing to help, quarreling at the drop of a hat—these are all the attributes of a difficult co-worker that no one wants on their team. Which type of co-worker do you want to be?

Tool Box Talk

"OK, gang, today I'm going to give a little pep talk about why we need to work as a team. This project we're preparing the site for, the Devlin Office Building, is already behind schedule because of some legal and financing matters I won't go into here. Bottom line is, we're the ones who have got to make the schedule back up. That means all of us, working together. We're gonna do it thoroughly, and we're gonna do it safe. First, we're going to mark where all the current utility lines are. We'll be getting reps in from water, electric, telephone companies; these are busy folks and they can't spend too much time here. So when they ask for your help, give it to them pronto. If those lines don't get marked properly, we disturb an underground electric line later on and that could cause a fire. Others of you will be helping to prepare access to the site; that all has to be finished before we can start clearing out the old drainage works and foundation. This is going to be a hot, dirty job where you'll be working closely with the heavy equipment folks. We do not want them sitting around, because they don't come cheap. So when they unearth something, for instance, we have to have enough hands on deck to remove that stuff so that the equipment operators can continue to do their jobs without damaging their equipment. We also have to fence off the site so that John and Jane Q. Public don't fall in some excavation and sue the tar out the company and so that any hot-fingered types in the neighborhood who are tempted to help themselves to piles or beams or…who knows, port-a-johns…won't be able to. There's much to be done, ladies and gents, so let's hop to it. Look out for each other, and help each other."

Here's a quick quiz that will allow you to apply what you've learned in this module. Select the best possible answer, given what you've learned.

1. Car troubles are making a co-worker, who lives just around the corner from you, late for work. To be a good team member, you should _____.

   a. advise your co-worker to get a more reliable car, and offer to help find one when the weekend comes around

   b. ask your boss to cut the co-worker some slack because that's the considerate thing to do

   c. try to get yourself transferred to another team because your co-worker's tardiness might affect your reputation for getting the job done on time

   d. volunteer to give your co-worker a lift to and from work until the car gets fixed

2. Your company hires an excellent craftworker for your team who speaks only a few words and phrases in English. To be a good team member, you should _____.

   a. mention to the boss that the team won't work as well because the worker, although talented, doesn't communicate well with the others, which will put the schedule at risk

   b. ignore the new worker as best you can, because the language barrier will make any attempt at talking a waste of time—time you don't have

   c. figure that you are both construction workers and can speak that language, and that any other language barrier can be overcome through effort if you learn a little of your co-worker's native language as that person learns more English

   d. ask the boss to put you where the team all speaks English because you can't follow instructions if someone giving it doesn't speak English as well as you do

3. A co-worker is having a week filled with problems: a fight at home, a car accident, and a sick dog that may die. These problems are affecting your teammate's concentration. You should _____.

   a. be sympathetic and offer a little help until things get better

   b. suggest that the co-worker read a book about stress later, but for now get back to work so that the team won't have to put in overtime to take up the co-worker's slack

   c. tell your co-worker that laughter is the best medicine, and then share a joke about a family arguing in a car and causing the driver to run over a dog and wreck the car

   d. tell your co-worker that no one wants to hear someone's personal problems because everyone has them, to quit ruining the team's morale, and to buck up

4. You come to work 15 minutes early each day, never stretch out your mealtimes or breaks, and are usually the last one to leave at the end of the shift. Your best course of action is to _____.

   a. make sure your supervisor knows you're the hardest-working member of the crew so that you'll get the biggest raise, because you deserve it

   b. tell your team members that their laziness is putting the whole project at risk and that they'd better pick up the pace

   c. recognize that everyone is busy and dedicated, not just you, and that your co-workers are probably working as hard, if not harder, than you are

   d. take notes on which crew members are coming in late, leaving early, or stretching out their breaks, then collect it all into a report and give it to your boss anonymously

5. Your company presents you with the Dedicated Employee Award at an annual breakfast meeting. You go up to accept the award and are asked to say a few words. You should _____.

   a. say "thank you" and then go straight back to your seat

   b. thank your team members and say that everyone on the crew deserves to share the award, because you're all teammates

   c. act like the award is not a big deal, because you don't want your team members to think that you're superior to them

   d. give a short speech explaining how your superior work habits set an example for the entire work crew

6. The electricians are due to begin wiring on Monday. They can't begin their job until you've finished yours, but you're running behind schedule, and it's already Friday. You don't think you can finish by the end of the day. You should _____.

   a. finish up work and go home, because it's the supervisor's job to worry about schedules

   b. talk to the electricians and say you're very sorry about the delay but that it's unavoidable

   c. ask your supervisor if you can work some overtime over the weekend so that the area is ready for the electricians on Monday

   d. cut a few corners to finish up everything by the end of the day and leave any problems for the electricians, who are all probably more experienced than you are anyway

7. A family argument kept you up late, and you wake up tired and annoyed. You should _____.

   a. call in sick, because you know you'll never be able to concentrate on your job and won't be worth much to the team

   b. plan on reporting to work around noon, because by then you'll be in a better mood and won't dampen everyone's spirits

   c. get to work on time, yell at a couple of the new workers, and slam a few doors to blow off steam and reduce stress

   d. get to work on time and put the argument out of your mind so you can concentrate on work

8. One of your teammates often comes back from breaks and meals later than everyone else. You should _____.

   a. get together with your teammates to play a pretty harsh joke that'll teach this character to be on time in the future

   b. expect that your boss has noticed or will notice the problem, and not worry about it any more

   c. tell your company's human resources department about the chronically tardy worker and how that's upsetting the team and putting the project at risk

   d. tell your teammate as sarcastically as possible how much you admire that person for putting one over on the boss and the company and the team, risking the project schedule and wasting money … and maybe the person will get the message

9. Your supervisor asks you to show a new trainee how to remove and replace a broken tile in a shower stall. The trainee can't seem to get the hang of it. You should _____.

   a. do the job for the trainee, because it'll save time and be less frustrating for both of you

   b. demonstrate everything slowly, step by step, and more than once if necessary

   c. tell the trainee to concentrate and to stop asking so many irrelevant questions

   d. sympathize with the trainee's difficulty in getting a handle on things, and then let the boss know later how hopeless it is to try and teach that person anything

10. You are having trouble getting a co-worker to understand how to operate a drill press. You spent much of last night drawing a chart with all the steps clearly written down, even color-coding the procedures. Today, however, the co-worker is still making mistakes. What's your best choice of action?

a. Ask your supervisor to find someone else to do the training because you obviously teach things in a way your co-worker can't understand.

b. Consider that maybe your co-worker is a hands-on learner, and then demonstrate how the drill press works.

c. Ask your co-worker that since you went to all the trouble to make such a great chart that any dummy could understand, what's the problem?

d. Ask, as gently as possible, whether the co-worker has some kind of learning disability.

Individual Activities

Activity 1: How Do You See Yourself As a Co-Worker?

The following chart lists behaviors and characteristics that tend to make people unpopular with their co-workers. Some people are greatly upset by some of this conduct; other people might not be bothered as much by that kind of behavior. Some people can't stand gossips; other people might rather be around a gossip than someone who is arrogant. None of these behaviors will win you a popularity contest, but it is the rare worker who never exhibits any of them. Rate yourself on how often you exhibit these behaviors or have these opinions. Rate yourself honestly.

		Often	Sometimes	Rarely	Never
1.	I don't like working with others in a team.	○	○	○	○
2.	I take myself very seriously.	○	○	○	○
3.	I lose my temper easily.	○	○	○	○
4.	I want to be the best at everything.	○	○	○	○
5.	I hold grudges.	○	○	○	○
6.	I show up late for work.	○	○	○	○
7.	It's not important to be considerate toward others.	○	○	○	○
8.	I take longer than allowed for meals or breaks.	○	○	○	○
9.	I like to gossip.	○	○	○	○
10.	I make other people's business my own.	○	○	○	○
11.	I'll report on my co-workers' behavior to my supervisor when I think it's necessary.	○	○	○	○
12.	I don't mind saying or doing things that upset my co-workers.	○	○	○	○
13.	I'll bad-mouth the company or the boss if they deserve it.	○	○	○	○
14.	I can't take a joke.	○	○	○	○
15.	I take credit for others' work.	○	○	○	○
16.	I tell everyone how talented and skillful I am.	○	○	○	○
17.	I am sarcastic.	○	○	○	○
18.	I don't listen to the ideas of others.	○	○	○	○
19.	I look down on people of other cultures or races.	○	○	○	○
20.	I don't want to pitch in and help.	○	○	○	○
21.	I horse around when others are trying to work.	○	○	○	○
22.	I work harder than most people do.	○	○	○	○
23.	I don't compliment people.	○	○	○	○
24.	I have a lot of resentment.	○	○	○	○
25.	I complain about my job, my co-workers, and my boss.	○	○	○	○
26.	I manipulate people.	○	○	○	○
27.	I look down on others because of their sex or age.	○	○	○	○

Now that you've filled out the checklist, for each behavior or opinion you checked as Sometimes or Often, come up with an action plan to help you improve that particular characteristic. A sample action plan to minimize the habit of gossiping is provided for you to use as a guide.

Action Plan for Improvement

Example

Sample Problem: *I like to gossip.*

Sample Action Plan:

1. Walk away when people start gossiping or bad-mouthing others. Or if I can't get away easily, change the subject.

2. Don't repeat gossip I hear.

3. Keep my nose out of other people's business.

4. Focus on my own work, not on what other people are doing.

Problem	Action Plan
I like to gossip.	1. Walk away when people start gossiping or bad-mouthing others. Or if I can't get away easily, change the subject. 2. Don't repeat gossip I hear. 3. Keep my nose out of other people's business. 4. Focus on my own work, not on what other people are doing.

Activity 2: Role-Playing Exercise—Who Would You Rather Work With?

In this activity, you will rate three workers based on what they tell you about themselves. Read the following three scripts and then, based on what you've read, rate them as co-workers using the following scale:

> **1** = I would really enjoy working with this person.
>
> **2** = This person would be OK to work with.
>
> **3** = I don't think I'd like working with this person.
>
> **4** = Working with this person would be terrible.

When all the scripts have been read, discuss what you liked and did not like about these co-workers. Then, as a class, come up with an action plan for each worker to improve teaming skills.

This can also be a group exercise, where you and a couple of fellow students read these scripts aloud to each other, and then discuss what you like and dislike about these people and whether you'd like to have them as teammates or not.

Co-worker #1 _____

I'm friendly and generous. When you work with the same people every day, you've got to be that way. I'm the one who breaks the ice with the rookies, makes 'em feel welcome. Me and the gang occasionally have a bit of sport with newbies, hide their tools, that kind of thing. Harmless stuff. It keeps up morale. With the crew we got, believe me, if I didn't keep things loose, some of these folks would waste no time throwing down on each other. Don't get me wrong, I'll help out the newbies, too. Today, for instance, I shared my lunch with one. Poor kid, first day on the job, had car trouble. So I offered to take a look at the car after work, said I'd fix it for free, too, if I could. Heh-heh. I had no intention of doing that, of course, but it just made the kid feel better. You could just tell. I've got a serious side, too. I got my share of good ideas. People always say I take forever to make my point, but that's just 'cause they aren't listening. Even the boss the other day said all my jawing gets in the way of other folks as they work. I thought, well if I talked that much, boss, I'd have told you go walk the short plank a long time ago. Told that to a couple of my work mates afterwards. Folks I could trust. On a crew, you gotta know who to trust. Usually it's pretty easy to find out. You just got to look at most folks, and there're some... you know, those people... well, you can't trust them. Not a one of them.

Rating: _____

Does this person have any qualities you'd like a co-worker to have? If so, what are they?

_____

_____

Could this person be a better co-worker? If so, how?

_____

_____

Co-worker #2 _____

I work really hard. I mean, the rest of the crew stroll in here seven-thirty, eight, I've been on the job since six-thirty. And I stay later than most of the others, too. They're so anxious to leave, impatient as a dog for dinner, and they leave their tools out. Who cleans it all up? Not the night watchman at the site, that's for sure! The other day, I was helping this new hire, kind of a clown, on how to dig trenches to set the water and sewer in. All day long I've been telling him, do this, do that, don't do that, get your head in the game, friendo! Then it's four-thirty, full half-hour before quitting, this character is AWOL. You see why then I'm not too keen on some of my co-workers. That doesn't go for everyone. One of them, Stevie, is always the first one to help you out. I put in a good word for Stevie with the boss the other day. I also put in something less of a good word about the goofball I'd been trying to teach. I'm close to the boss, who's kind of like my equal. I'll tell you why. Other day, I was hauling some materials out to Esperania with a co-worker who just the day before had been diagnosed with leukemia. I tried to commiserate a bit, but what can you say? Well, what you can do is cut the person a break, and that's what I tried to do. I did all the unloading and loading, told my co-worker not to worry about it. We were a little late in getting back, and the boss kind of chewed me out about it. I said it was all my fault, I just wasn't working fast enough, I said. It was the least I could do for my co-worker. I mean, being told you have cancer. Boy, that's rough. Anyway, the reason why I got away with it is because I'm close to the boss.

Rating: _____

Does this person have any qualities you'd like a co-worker to have? If so, what are they?

_____

_____

Could this person be a better co-worker? If so, how?

_____

_____

Co-worker #3: _____

I've been with the company for about 10 years. I've seen them come and go. I know just about everything there is to know, and what I don't know probably isn't worth knowing. I can work pretty well on my own, don't need to be part of a "team." I don't need to be slowed down or bothered by some neophyte every time I reach back on my work belt for my tape measure. The other day, a real scorcher, two of the people I work with were arguing like mad about whose turn it was to take away the haz-mats. Neither one of them wanted to put on the heavy gear in the hot weather. Heck, I said, I'll do it, just to get away from you lot. Cleared it with the boss, who'd probably rather me do it anyway, knowing how thorough I am. I don't hold with gossip or with talking behind folks' backs. It's not polite, and it's cowardly. I don't hold with moaning about your problems, either. I got mine, I don't need to hear about yours, thank you. Doesn't make me unfeeling, but a worksite isn't the place to get all emotional. That's why I don't buy certain folks working there … you know, that half of the population. Just not the place for them. I don't care if you've got certifications galore, probably got them through some kind of favoritism anyhow. Emotional types can't cut it at a construction site, and that's that.

Rating: _____

Does this person have any qualities you'd like a co-worker to have? If so, what are they?

_____

_____

Could this person be a better co-worker? If so, how?

_____

_____

Activity 3: Building a Work Crew, and Being Part of One

This is a two-part activity. The first part will give you some idea of what goes into choosing a crew that can work well together. You might not get a chance to do this much in the real world, at least while you're starting out, but it will give you an understanding of how different personalities can affect the success of a crew, the effectiveness of a team, and the outcome of the project. You'll gain even more understanding from the second part of this activity, which gives you and your classmates an opportunity to see how this team might work together, by each taking a role of one of the five team members and acting out the part.

You'll be assigned to teams of five. Read the construction project. For the first part of the activity, you and your teammates will choose what your team thinks is the best mix of a carpenter, plumber, electrician, painter, and cabinetmaker for the project. You'll choose from a list of candidates in each of these occupations, which is provided after the project description. When discussing who should work on the project, you should consider not only the individuals' skills, but also their personality traits that may affect their ability to work well with others.

Team Members

1._____

2._____

3._____

4._____

5._____

Construction project: Your company is working on a new terminal for a regional airport. The airport authority—the customer and eventual operator of the airport—has ordered some modifications to the baby-changing stations in the five men's restrooms the new terminal will have. The new plans call for the addition of a drop-in sink and a cabinet to hold the sink and supplies. The new plans also call for the sink drain to go where the current plans had an electrical conduit, so both the plumbing and electric feeds will need to be reconfigured, and it will be delicate work. In addition, the cabinet will require a special countertop design so that when babies are placed on it they won't roll off by accident. Two new walls—a full wall and a knee wall—must be built and finished for each station, but adding these walls will require close cooperation with the plumber and electrician because the space for the redesign is limited. Because the restrooms are more or less finished except for the baby-changing stations, all this work will have to be done neatly so that the rest of the bathroom is not affected. The overall airport project is behind schedule, and everyone is under pressure to get things done. With a good work crew, you're confident this particular task can be completed on time. Based on the work to be done and the circumstances, choose five team members—one from each trade—from among the following individuals:

Carpenters

Pete: An experienced and top-notch carpenter, Pete has a reputation for being moody and difficult to work with, possibly because he's going through a nasty divorce.

Janice: An outstanding carpenter, known for her fine work. Lately, Janice has been getting in late and leaving early, and has started missing deadlines.

Chanelle: A pretty good carpenter, given she has about 4 months of experience. Chanelle works fast (sometimes too fast), always wants to learn something new, and enjoys her job.

Plumbers

Mike: A top-notch plumber who's also skilled in electrical work, Mike has a big ego and can make co-workers feel inadequate, especially those who don't know as much as he does. Mike gets things done and doesn't hesitate to tell you all about it afterward.

Ahmed: A good, helpful plumber who has some carpentry skills as well. Last week, Ahmed helped some carpenters at another part of the site when they were short a member, although he fell behind schedule on his regular job as a result.

Jorge: A good plumber who just became a journeyman, Jorge speaks English as a second language. He has a few problems every now and then with words and phrases, which affect his overall eagerness to learn and help.

Electricians

K.C.: A top-notch electrician who is efficient and neat, K.C. has a reputation for having problems working with people from other ethnic backgrounds. K.C. has never missed a deadline and is good at moving things along.

Zack: A good electrician who also works pretty fast, Zack likes to tell jokes, especially about people of other backgrounds than his. On the other hand, Zack is always willing to show someone less experienced how something is done. Zack likes long lunches.

Johnni: A trainee who has just become a journeyman, Johnni has a knack for getting everyone to work together as a team. Sometimes Johnni, who works very deliberately, has trouble meeting deadlines and can get easily upset by criticism.

Painters

Kwanlee: A skilled, neat, and quick-working painter, Kwanlee is easygoing but tends to waste other people's time gossiping about co-workers and has something of a reputation for spreading rumors.

P.J.: A skilled painter who on occasion can be a bit sloppy, P.J. loves coming to work, and most people enjoy his upbeat personality. P.J. sometimes leaves a messy work area behind at the end of the day.

C.C.: A skilled and neat painter, C.C. does not like to be rushed at the job and prefers to work quietly. C.C. is cordial to co-workers but gets irked if someone tells her to speed up or gets involved in her business.

Cabinetmakers

Nadia: An able craftworker who also has some plumbing experience, Nadia is quick-witted and funny but tends to interrupt co-workers, finish their sentences for them, and tell them what they're doing wrong.

Goran: Fast, neat, and thorough, Goran feels he works harder than everyone else and so should be paid more. He has a reputation for having a bit of an attitude problem and for letting supervisors know when people leave early or come in late.

Ben: A great craftworker with a lot of experience, Ben is cordial and polite with co-workers, but he doesn't think that women belong in construction. He'll compliment a male co-worker readily, but will rarely, if ever, do so with a woman.

For the second part of this activity, each of your teammates will take one of the roles (or your instructor may assign the roles). Discuss how the strengths of the five people you've chosen will help to get the project done. Discuss how the shortcomings of the five people you've chosen might delay the project or have a negative influence on the team. A few discussion questions have been provided, but you and your teammates can discuss others as well.

Discussion Questions

1. What effect does it have on productivity if some team members don't like working with others, for whatever reason? Are some reasons more damaging to the team than others?

2. What negative characteristics of the team members you've picked are the greatest risk to the effectiveness of the team? What positive qualities of the team members you've picked are the most beneficial to the team?

3. Among the five people you've picked, which team members are most likely to come into conflict, and for what reasons? Are there other members of your team who have qualities that could help smooth over those conflicts?

4. Are the shortcomings of some team members balanced by the strengths of others?

Diversity in the Workplace

"(People) may be said to resemble not the bricks of which a house is built, but the pieces of a picture puzzle, each differing in shape, but matching the rest, and thus bringing out the picture."

– Felix Adler,
philosopher and
social reformer

The construction workforce includes people of every age, race, color, gender, sexual orientation, religion, ability, and ethnic background. These traits are often referred to as demographics. Consider these charts on the demographic makeup of craftworkers and construction laborers in the United States in 2007.

Percentage of Craftworkers by Race (2007)

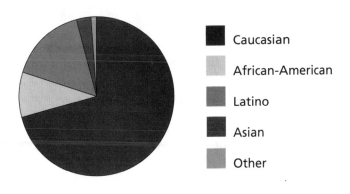

Caucasian

African-American

Latino

Asian

Other

Percentage of Laborers by Gender (2007)

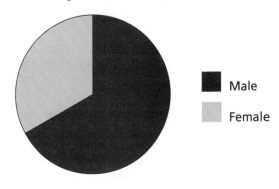

Male

Female

Percentage of Laborers by Race (2007)

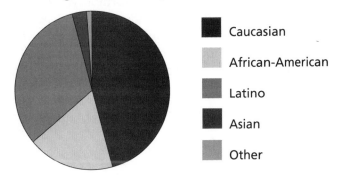

Caucasian

African-American

Latino

Asian

Other

This graph illustrates how the demographic makeup of the construction industry has changed during the decade of the 2000s.

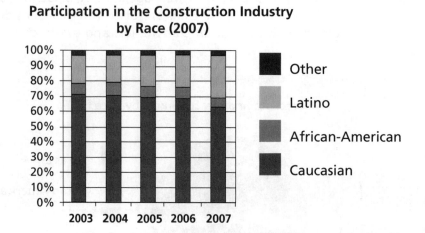

This graph shows the growth of the number of female employees in the construction industry over the past few decades.

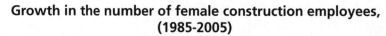

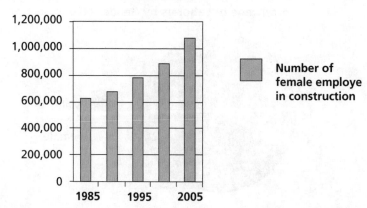

These statistics indicate that the construction workplace is diverse and becoming even more so. To be successful in your career, it's essential that you are comfortable working along with people who aren't from your background or who don't have the same demographic attributes as you do. Regardless of the differences in age, gender, race, ethnic background, or any other demographic among your co-workers, you all have something profound in common: a profession and the mutual desire to do well in it. This module discusses workplace diversity and shows you how to succeed in a diverse workplace.

What is diversity?

Diversity means *variety*, not just differences. A diverse workplace is one with a wide assortment of people: men and women of varied ages, races, national origins, ethnic backgrounds, sexual orientation, religions, and abilities.

Diversity can enrich our lives, bringing a wealth of different perspectives and experiences to our relationships and workplaces. Diversity can also cause tension, especially in situations where people engage in negative behaviors such as stereotyping, being prejudiced, and discrimination. These three things are toxic influences at any worksite, and are barriers to success in the construction industry or anywhere else.

Working with diverse co-workers

Consider all the ways people differ from one another: by age, race, gender or sexual orientation, the language they speak, the nation they're from, their religious beliefs, cultural practices, and personal experiences. At some level, we're all fundamentally unique. You could say that diversity is one of the few things all of us have in common.

As a member of a diverse work crew, you'll be spending several hours a day with co-workers who will likely be different from you in many ways. You'll be depending on these people to help get the job done, and you'll sometimes have to trust them with your safety. Showing them the respect to which they're entitled is the right thing to do, and it is vital to your job.

Throughout this book, we've talked about the need to respect your supervisors and co-workers. That means everyone, not just people who come from your background. Those who say, "I can only work with people who are like me" won't last long at any worksite.

During your career in construction, you may work at jobsites where you share the demographic attributes of most of your co-workers, but you may work at jobsites where you don't. If you are comfortable working with all kinds of people, it shouldn't matter. Here are some tips on how to act in a diverse work environment.

Look for common ground. At your job, you'll be working with people who have different backgrounds, beliefs, or cultures. You may think you have nothing in common with them, but you do; you're working on the same project, and that means you have vital goals in common:

- You want to do quality work.

- You want to finish the job on time and within budget.

- You want to stay safe on the job and do things properly.

- You want to make a good living.

Once you realize how important these goals are, you'll realize how absurd it is to concern yourself with whether someone has a different background or appearance than you do.

Respect cultural differences. People from different cultures might approach life and its challenges in entirely different ways. For instance, people from Asia tend to solve problems in groups, while people who were born in the United States often favor the individualist approach. Be aware that different religions require believers to act in certain ways at certain times of the day or year, and what's acceptable to members of one culture may make members of another culture uneasy. Stop and think how you would feel if someone said you were strange because of your beliefs or practices.

Think of your co-workers as individuals.

It's discourteous to consider a person as nothing more than an extension of some group. It is inaccurate to presume that all members of a group think or act alike. Everyone wants to be:

- Treated with respect.
- Treated as individuals.
- Valued for skills, talents, and experiences.
- Accepted as members of the work crew.

Are you a newcomer to this country?

Behavior considered in some cultures to be rude, hurtful, or embarrassing may not be considered that way in the United States. There are some basic attitudes and beliefs common among Americans that might help to guide a newcomer to this country on how to act on the job:

- People can and should control their own destiny and behavior, and therefore people are responsible for their own situations in life.
- Individualism is important. (Americans value their privacy.)
- Time is a resource, like water or oil, and therefore shouldn't be wasted.
- People have equal opportunities.
- You earn respect through your achievements; no one is born better than anyone else is.
- Change is usually good, because it's associated with progress, improvement, and growth.
- Honest argument is often better than polite but insincere agreement, and criticism is acceptable as long as it's given respectfully and in a friendly manner.

If you are a newcomer to the United States and unsure on how to act on the job, it's best if you avoid making assumptions about what is considered appropriate conduct. If you are confused about how to handle a certain situation, you can ask a co-worker or local resident about it.

Potential problems to watch for

As you work with others, you should be careful to avoid three offensive habits that will damage your working relationships and career. These habits are stereotyping, being prejudiced, and discrimination.

Stereotyping

A stereotype is an unfair and often negative image about all members of a group. Two examples of stereotypes are that all construction workers are uneducated, and that athletes are all muscle and no brain. When you generalize about groups of people, it is difficult to recognize the qualities and skills of individuals. Stereotypes are often the basis for prejudice, which is discussed below. To avoid stereotyping others, ask yourself the following questions:

- **Am I being fair?** Am I generalizing about the abilities or habits of others based on their race, color, gender, sexual orientation, religion, national origin, age, or disability? How would I like it if others made such assumptions about me and ignored me as a person?

- **Do I judge a group of people based on what happened in my past?** Did I have a bad experience involving someone from a certain racial or ethnic group, and have I held that against everyone from that group ever since? How many rewarding relationships have I missed as a result?

- **Did I learn this stereotype?** Did members of my family believe in this stereotype? Did I learn it from them? Did I start believing in this stereotype so that I would fit in with others? Am I going to pass this stereotype along to my family the way it was passed along to me?

Prejudice

Prejudice occurs when someone judges others based on their group membership, without knowing anything about them as individuals. A prejudiced attitude is usually, although not always, unfavorable or hostile, and it's always unreasonable. Prejudices can be based on practically any characteristic: race, sex, ethnic or national origin, age, religion, sexual orientation, or disability. Actions based on prejudice will hinder your job performance and could lead to losing your job. To avoid or overcome prejudice, follow these guidelines:

- When you form an opinion about someone, ask yourself what information went into making that opinion. Did you use stereotypes? Or did you consider the person as an individual and not just as a member of some group?

- Take time to get to know your co-workers as individuals. Ask them about their friends, their families, and what they want out of life. You'll find that you have more in common with them than you may have first thought.

- Be open-minded and don't be quick to pass judgment on people; everyone deserves a chance.

- Remember that variety is the spice of life. A world in which everyone had the same background as you wouldn't be very interesting.

Discrimination

Discrimination means treating others differently because they belong to a distinct group or category. Discrimination in employment happens when workers are denied job opportunities or assignments, raises, or promotions because of their group membership. Specific national laws prohibit employers from discriminating based on race, color, gender, religion, national origin, disability, and age. Some states and cities forbid discrimination based on sexual orientation, marital and family status, and political affiliation.

Discrimination is not only illegal but, like stereotyping and prejudice, is unfair and mean-spirited. All employees are entitled to equal treatment. You do not have to be a supervisor to be guilty of discrimination. To avoid discriminating against others:

- Don't form opinions of others using prejudiced attitudes.

- Think of your co-workers as individuals.

- Treat your co-workers with respect, and remember the golden rule.

Summary

Today's diverse workplaces demand that those who work there are accepting of all people regardless of their age, gender, color, race, religion, national origin, disability status, or sexual orientation. Stereotyping, being prejudiced, or discriminating against others are all unacceptable behaviors. That kind of conduct will damage your work relationships and may ruin your construction career. Remember that a diverse workplace means variety, and diversity enriches our lives. If it occasionally leads to tension, that tension can be defused by treating each person as an individual, respecting cultural differences, and looking for common ground.

A Note On Generation Diversity

Which of these events was a defining moment to you: the moon landing, the fall of the Berlin Wall, or the O.J. Simpson trial? Which of these deaths affected you the most: Janis Joplin's, John Lennon's, or Tupac Shakur's? Your answers could say a lot about which generation you belong to.

There are generally three different generations in the workplace today: the Baby Boomers, who were born between World War II and the early 1960s; Generation Xers, born between the mid-1965 and 1976; and the Millennials, who were born between 1977 and 1998. You'll work with members of all three generations.

Different generations have lived through different times and have had different influences. Remember, however, that Baby Boomer, GenXer, and Millennial are labels that refer to when people were born; they aren't signposts that automatically tell you how a person thinks or what a person believes. It would be an unwise generalization to assume that members of the same generation all have the same opinions or character traits.

Nevertheless, there are cases when disagreements can be traced to differing generational outlooks. If you understand and respect these differences in outlook, if you look upon it as generational diversity that adds valuable perspectives to your jobsite, you'll get along better with people of all ages. If you're in a conflict with a co-worker and think the issue is related to generational outlook, here are some ideas on how to resolve the situation:

- Think back on what you've said; a comment that might not upset someone in your generation might be considered offensive by a member of another. Try to avoid miscommunication in the future.

- Look for common ground. You may have different ideas from your older or younger co-worker on how to do something, but you share the desire to do a good job.

- You may not agree with the values of those in different generations, but you must respect those values. Remember, they have been derived from experiences as valid as yours.

- Remember that we all go through the same stages in life. If you're not young now, you were; and if you're young, you get older every day .

On-the-Job Quiz

Here's a quick quiz that allows you to apply what you've learned in this module. Select the best possible answer.

1. One of your co-workers must leave early one day to celebrate Passover and has asked your help to finish a task. (Passover requires those of the Jewish faith to stop work by sundown so that they may worship in the traditional way.) What is your best course of action?

   a. Give your co-worker a hand and help finish the work.

   b. Complain to your supervisor that your co-worker is trying to cheat the company out of time.

   c. Convert to Judaism, so that you can leave early too.

   d. Explain that while you respect your co-worker's religion, you don't share the belief and can't help finish the work.

2. You are working with a person whose religion, Islam, requires its practitioners to kneel and pray five times a day. You've just started a new project that will last two weeks. What is your best course of action?

   a. Tell your co-worker that the jobsite is no place for prayer.

   b. Joke about how the co-worker's knees must get awfully bruised from all that kneeling.

   c. Say nothing at all—just count the days until the two weeks are up.

   d. Say nothing to your co-worker, but if the praying interferes with getting your work done, ask your supervisor what you should do to keep on schedule.

3. You're a young African-American woman, and you've just started a new job. Your team members are an Asian-American man, a young Latino woman, and an older white man. What is your best course of action?

   a. Work with them but don't bother getting to know them.

   b. Introduce yourself to your three co-workers by making jokes about the lesbians you used to work with.

   c. Tell your boss that you'd feel more comfortable working with people of your own age and background, because you'll work more quickly and safely around people who are like you.

   d. Introduce yourself to your three co-workers and get to know them as individuals.

4. You're male, and a female electrician has just joined your team. When you meet her, what is your best course of action?

   a. Hug her and say, "Welcome aboard, sweetie."

   b. Welcome her to the job and introduce her to the other people in the work crew if she hasn't met them already.

   c. Warn her that the work can be very hard physically.

   d. Nod hello but don't say anything to her because she might accuse you of harassment.

5. One of your male co-workers, a military veteran, has been with the company for 15 years. The younger workers make a lot of jokes about older people. They often use terms such as *geezer* and *old-timer*. They say that their attitude and remarks are all in good fun. If asked to comment on this behavior, you should say, _____

   a. "As long as he's OK with it, it's totally fine."

   b. "It's mean-spirited and disrespectful to make jokes about a worker's age, no matter what it is."

   c. "On a construction site, you have to know how to take a joke."

   d. "The guy is a former Marine, so he's tough enough to let all that slide."

6. Your supervisor has recently hired a Hispanic worker to help out around the jobsite. At present, your new co-worker speaks only a little English. You should assume that your co-worker _____.

   a. won't speak English much better any time soon because he is probably too lazy to take any English-language courses

   b. is probably in the United States illegally

   c. has probably come to the United States to seek a better life and should be admired

   d. could use some help with the language, so you should take time during the work day to give your co-worker a few informal English lessons

7. A worker with a prosthetic (artificial) arm and hand is assigned to your team. You have never been around anyone with this type of disability before. What is your best course of action?

   a. Show pity for your co-worker but assume that you will have to do extra work.

   b. Tell your co-worker all about your grandmother who has an artificial hip and can't get up and down stairs any longer.

   c. Ask your co-worker to tell you how the disability happened and ask if you can borrow the prosthetic device to better understand the person's disability.

   d. Get to know your co-worker as an individual and make accommodations for the disability if you think it's necessary.

8. A co-worker doesn't understand the meaning of the word stereotype, so you provide a few examples. What is *not* an example of stereotyping?

   a. All workers want to earn a good salary in a safe workplace.

   b. All male construction workers are beer drinkers and whistle at women walking by.

   c. Women never last long at a construction job because they end up getting pregnant and have to quit.

   d. Never put an African-American and a white boy from the South on the same team; they just won't get along.

9. You have a good friend at work who often tells jokes about others from different backgrounds. What is your best course of action?

   a. Tell your friend to keep it down because if the wrong person hears the joke, you'll both be in trouble.

   b. Tell your friend you know a better joke about another group of people—one that isn't represented at the jobsite.

   c. Say that these jokes are OK when you're not at the worksite but could get in the way of job if someone heard them at work.

   d. Explain to your friend that the jokes are stereotypical, unfair, and insulting.

10. If your supervisor _____, your company could be fined for discrimination.

    a. reprimands an African-American worker for coming in late several times in one week

    b. reprimands an Asian-American worker who is not wearing safety gear

    c. refuses to promote a female worker because women don't belong in construction

    d. fires a disabled worker who keeps showing up to the job drunk

Group Activities

Activity 1: Walk Apart – Walk Together

Ask for two volunteers to come forward and stand with their backs together. The rest of the class is asked to call out things about these two volunteers that are different. Differences sometimes push us apart. As each difference is called, the volunteers take one step apart. When they reach the end of available space, have them turn and face each other. Now the class is asked to call out things that are similar/alike about the volunteers. As each similarity is called out, the volunteers take one step toward each other.

Note that most times the differences are things we can see: hair/skin color, wearing glasses or not, different type shoe, and so on.

The similarities are often things that others perceive: both are in the class, they do the same type of work, etc. Of course, there may be some physical characteristics that are similar, too.

Activity 2: Cultural Scavenger Hunt

As you chat with classmates, find people who have had the following experience(s). Have them sign their name or initials in the appropriate blank. Try to talk to everyone in the room.

_____ 1. Knows a folk dance.

_____ 2. Has been to an American Indian pow wow.

_____ 3. Has cooked or eaten ethnic food in the last week.

_____ 4. Can say "hello" (or similar greeting) in four languages.

_____ 5. Has sat under a palm tree.

_____ 6. Has attended a religious service of a religion other than their own.

_____ 7. Has attended a Kwanzaa celebration, or knows what Kwanzaa is.

_____ 8. Has relatives or ancestors who came through Ellis Island.

_____ 9. Plays a musical instrument.

_____ 10. Has had to utilize crutches, a wheelchair, a cane, or has worn a cast.

_____ 11. Can name four different kinds of breads from other cultures.

_____ 12. Has seen a Spike Lee movie.

_____ 13. Is bilingual, or has relatives who speak a language other than English.

_____ 14. Knows some American Sign Language.

_____ 15. Likes to do jigsaw puzzles.

_____ 16. Has studied a foreign language.

_____ 17. Has been a pen pal.

_____ 18. Has attended a Las Posadas celebration, or knows what Las Posadas is.

Activity 3: Learning to Have Empathy

One of the best ways to understand other people is to put yourself in their place. For this activity, you will work in groups of three or four. Each member of your team will read the five discussion questions below, and then discuss them as a group. Your goal is to think about times when you felt different from other people—times when you felt you didn't fit in. Here are some examples:

- You're in a restaurant where everyone is dressed up and you're wearing jeans and a T-shirt.

- You're the only man in a room full of women, or the only woman in a room full of men.

- You're the only person of color in a room full of whites, or the only white person in the room.

- You're the only one without a date at a party.

- You're in a waiting room before a job interview, and you notice everyone else who is applying for that job is considerably younger than you are, or considerably older than you are.

Your goal is to think about a time when you felt different from other people, a time when you felt you didn't fit in.

Team Members

1. _____

2. _____

3. _____

4. _____

Discussion Questions

1. Describe the situation. Who was there? When did it occur? What exactly happened?

2. How did the other people treat you? Did they ignore you? Did they stare at you? Did they lean over to the person next to you and whisper a comment, while looking at you indirectly? Did they come right out and make fun of you? Or did they try to include you and make you feel comfortable, and ask questions to learn about you as an individual?

3. How did you feel at the time? How did you react? Do you think you handled the situation well? If not, what would you do differently if you found yourself in the same situation again?

4. Did the situation help you understand yourself better?

5. How can you learn from your experience? How can this experience help you work better with people from different age groups and other cultures or backgrounds?

Activity 4: The Acceptance Profile

In this activity, you and your classmates will develop three profiles. Look at the following list and put an X next to words that you think describe a tolerant person. Put a Y next to words that you think describe an intolerant person. Put a Z next to words that you think could describe either a tolerant or an intolerant person. After creating your tolerance profile, discuss the questions that follow.

Friendly	Quiet
Judgmental	Brave
Open minded	Imaginative
Closed minded	Pushy
Suspicious	Patient
Willing to listen	Bitter
Curious	Inconsiderate
Outgoing	Upbeat
Self-confident	Outwardly focused
Frightened	Stubborn
Compassionate	Talkative
Jealous	Reasonable
Impatient	Depressed
Bossy	Even tempered
Self-focused	Helpful
Unhappy	Unreasonable
Thin skinned	Perfectionist
Harsh	Unwilling to listen
Ruthless	Mean spirited
Manipulative	Gentle

Team Members

1. _____

2. _____

3. _____

4. _____

Discussion Questions

1. Which of these characteristics do I admire most in a person?

2. Which of these characteristics do I value most in myself?

3. What things can I do to achieve the characteristics I most admire and value?

4. Did all of the team members agree on the qualities that characterize a tolerant person? What qualities characterize an intolerant person? What qualities could be used to characterize either a tolerant or an intolerant person? Discuss qualities on which you disagreed. Does a particular quality always indicate that person is tolerant or intolerant?

Communication Skills I: Listening and Speaking

"What we've got here is a failure to communicate."

– Paul Newman, in the 1967 film,
Cool Hand Luke

Introduction

What's the most important skill you need as a construction professional? Many construction workers would say, "Knowing the right way to use tools and equipment." It's true that you need good technical skills to succeed. But how do you learn those skills in the first place? You learn them by listening closely in class or when someone is tutoring you; by reading textbooks, instructions, or manuals carefully; and by knowing what questions to ask as you are learning.

Basic communication skills—listening, speaking, reading, and written communication—are fundamental to just about any task in life, and that certainly includes your job in construction. Here are examples of why you need each of these four skills in construction work:

- **Listening.** Your supervisor tells you where to set up safety barriers, but you don't listen carefully, and as a result, you miss one spot. That's where your co-worker falls and gets seriously injured.

- **Speaking.** You have only a couple of hours to train two new workers to do a task, but as you're telling them what to do, you mumble, use terms they don't understand, and don't answer their questions clearly. As a result, your co-workers do the job incorrectly, and the entire team must put in overtime to correct the mistake. This costs the company money, sets the schedule back, and results in a lot of cranky teammates.

- **Reading.** Your supervisor tells you to read the manufacturer's basic operating and safety instructions for a new precision-built spindle before you use it. You don't really understand the instructions, but you don't want to bother the boss. You go ahead and use the unit as you think it's supposed to be used, and you damage the equipment to the point it can't be repaired. As a result, the company has to buy a new unit.

- **Written communication.** Your boss asks you to write up a material takeoff (a supply list) for a project. You are in a hurry, so you don't write down the specifications clearly, and you don't check what you've written. As a result, the supplier delivers something other than what your boss wanted, and the company has to reorder the supplies. This costs a lot of money that could have been saved if you'd written down the order more clearly in the first place.

Communication on the worksite directly influences safety, schedules, budgets, and morale. Improving your communication skills will make you a more valuable employee. This module focuses on listening and speaking skills. *Communication Skills II* focuses on reading and written communication skills.

How communication works

Before we discuss how you can improve your listening and speaking skills, we should first understand how communication works. *Figure 1* is a simple diagram of the process, which is applicable to all four communication skills: listening, speaking, reading, and written communication. There are two basic steps to communication:

Step 1: A spoken or written message is sent from one person (the sender) to another (the receiver) through a variety of communication channels: phone, voice mail, email, text message, two-way radio, face-to-face conversation or meeting, or a written note or memo.

Figure 6-1. The Communication Process

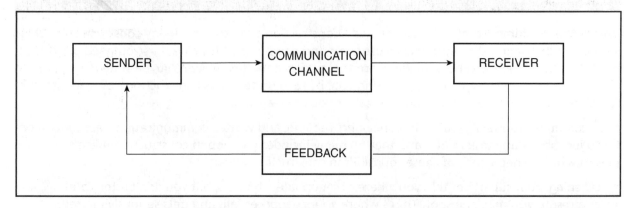

Step 2: The receiver gets the message by listening to the sender or reading what the sender has written, and figures out what the sender is trying to communicate. If anything is not clear, the receiver asks the sender for more information either in real time (in a phone conversation or a face-to-face meeting) or by responding to the sender via an email or a note or a voice mail message.

This process is called two-way communication, and both the sender and receiver must be good communicators for it to be effective. Two-way communication seems simple, but it is difficult to practice and often ineffective.

There are a lot of potential obstacles to communication. Some people refer to these obstacles as noise. There are a number of circumstances that can obstruct effective communication:

- The sender uses work-related words or jargon that the receiver does not understand.

- The sender mutters in conversation or is otherwise unclear, or the sender does not speak clearly when leaving a voice mail message or when talking over a cell phone or two-way radio.

- The sender's written message is disorganized or contains mistakes, or the sender's handwriting is illegible.

- The sender's message is vague or does not have enough detail, whether that message is written or spoken.

- The sender does not get to the point when speaking or writing.

- The receiver is tired, distracted, or just not paying attention.

- The receiver has poor listening or reading skills.

- The receiver cannot hear the message or concentrate on understanding the message because of the actual noise that is common to a construction site: for example, coming from pneumatic hammers, earthmoving equipment, jackhammers, or other loud equipment.

- There is a mechanical problem with the equipment used as the communication channel: for example, there is static on a two-way phone or cell phone.

Effective communication can also be hindered by different styles of listening, reading, writing, or speaking on the part of both sender and receiver. The receiver may be unfamiliar with the sender's style of writing, or the sender may be more familiar with how text messaging works than the receiver is. In your construction career, you'll note that different supervisors, as well as your various co-workers, may communicate most effectively in different ways or will prefer one method of communication to another. That's why it's important that you have solid communication skills across the board.

Becoming an effective listener

Many people think that as a listener, you just sit or stand there while someone is talking. But effective listening means understanding what has been said, which means the listener must take an active role for effective two-way communication to occur. You don't listen to safety instructions the same way you listen to music. Here are tips to help you become an effective listener.

> "It was impossible to get a conversation going; everyone was talking too much."
>
> – Yogi Berra

Understand the consequences of not listening. In the introduction, you read an example of what can happen on the jobsite if you don't listen carefully. Can you recall a time when you failed to listen carefully, which led to problems on the job or in your personal life?

Stay focused on the speaker. Don't allow your mind to wander. This can be tough, especially when you are tired or hungry. To stay focused, it helps to make and keep eye contact with the person who is speaking. Don't stare off into space or fiddle with equipment. Don't think of other things; devote your attention to the speaker. Don't yawn, slouch, or shake your head; these may be signs that you are not listening closely enough. They might also indicate to the speaker that you are not paying much attention, which could aggravate the person talking to you.

Keep an open mind. When someone starts to say something you already know or disagree with, you need to keep listening regardless. For instance, you may think you know all about a procedure the boss is describing, but the point might be that the procedure has been modified from how it's usually done. You might disagree with what a speaker is saying, but the speaker might be right and you might be wrong. Remember, on a construction site, all communication is important—especially with your supervisor—and it needs to be heard and understood.

Take an active role to get the information you need. If you don't understand something, ask the speaker to clarify what was said. If some information seems incorrect to you, ask the speaker whether that information is correct; the speaker might have misspoken, or might not be good at spoken communication. If you can't hear what the speaker is saying, ask if you both can move to a quieter location, or ask the speaker to talk a bit louder.

Pay attention to the details. Construction work involves a lot of specifics: sizes, locations, and times. Listen closely to any communication dealing with numbers or figures, such as measurements, amounts, or driving directions. Get all these figures and numbers straight, and don't be reluctant to ask about a detail that seems wrong or unclear. Did that supplier on the phone tell you to pick up the sheet metal at *615* Wyant Street or *650* Wyant Street? Was that *Wyant* Street or *Wyatt* Street? If you've made the pickup before at 615 Wyant and the person on the phone is telling you to go to 650 Wyatt, you should ask for clarification. Ask the speaker to spell out the name of the street if you're unfamiliar with it.

Take notes. When appropriate or possible, jot down details as the speaker is talking. It's a good habit to have even if it's not always practical to take notes at a jobsite. Add a little pad of paper and a pen to the tools you carry with you every day. Supervisors appreciate it when you take careful notes; it demonstrates that you are taking responsibility for your work and approaching the job as a professional.

Don't guess. Sometimes an instruction may not be specific enough. For instance, your supervisor may tell you to change the tolerance on a machine part. Does that mean you're supposed to increase the tolerance or decrease it? And by how much? Don't try to guess the answer. Ask for clarification.

Ask what words mean. Since you're new on the job, you'll likely hear some words and expressions you haven't heard before. You'll hear people use abbreviations like *OC* or *MSDS*. Then there are common words like *specs, mud*, and *rock*, which have meanings in the construction industry that are entirely different from how the words are commonly used. How would you interpret the following instructions?

- Go to the office and pick up my specs.

- Put a coat of mud on these sections of drywall.

- Rock this room this afternoon.

- The studs should be 16 inches OC.

- Make sure you read the MSDS before you open that can of solvent.

Don't be embarrassed if you don't know the meaning of something. You are going to hear a lot of things you don't understand at first. Your boss knows this and expects you to ask questions. In construction, an unasked question can be hazardous to your health. Stay safe by staying informed.

Terminology

So what is meant by *specs, mud*, and *rocking* a room? Your boss is not asking you to pick up a pair of glasses, dirty up a wall, or play loud music. What about *OC, red-tag*, and *MSDS*? In case you don't already know, here is what these terms mean in construction:

Specs: An abbreviation for *job specifications*, which are the detailed descriptions of the job, including the materials needed to complete it.

Mud: In construction, this word refers to joint compound, which is the material used to fill in the thin space (the joint) between pieces of drywall installed next to one another.

Rock: A short form of *sheetrock*, which is the material used to build interior walls. Sheetrock is basically a brand name for drywall (in the same way that Kleenex is basically a brand name for tissue paper).

OC: An abbreviation for *on center*, which is the measurement from the center of one stud (a wall or ceiling support) to the center of the next.

Red-Tag: This is when someone places a red warning tag on equipment. The tag alerts other workers to an unsafe condition.

MSDS: An abbreviation for a *material safety data sheet*. An MSDS is a complete description of a product, including any safety hazards and recommendations for safe use. For example, an MSDS for a solvent may recommend that the person applying it wear a respirator.

Paraphrase back to the speaker what you've heard. When you paraphrase, you listen to what people say to you and then say it back to them in your own words. Suppose your boss says, "Red-tag this scaffold but set a watch on it first. It's pretty shaky." You listen and then say, "OK, I'll fill out the tag and tie it on the scaffold, and I'll get C.M. to keep everyone off 'til I finish." When your supervisor nods, you'll know you got the message right.

Barriers to listening

As we've seen so far, to become an effective listener involves developing a few important skills. But mastering those skills isn't all there is to effective listening. There's always the possibility that communication could be complicated or even thwarted by obstacles such as what kind of mood you're in while you're listening; how the speaker is presenting the subject; external circumstances; distractions; or your ego. These obstacles can undermine your grasp of a message no matter how good your listening skills normally are:

- **Emotion.** When people are angry or upset, they usually stop listening. If you find you are getting mad while someone is speaking to you, or if you are being upset by what the person is saying, try counting to ten or ask the speaker to excuse you for a minute. Go get a drink of water, or take a couple of deep breaths. But avoid getting into an argument.

- **Boredom.** Maybe you've heard it before, and don't see a reason to listen again. Or maybe someone is just droning on and on and you're wondering when the speaker is ever going to stop. There's no easy way to overcome boredom. You just have to make yourself stay focused. Keep in mind that the speaker has important information you need to hear. It's considerate and respectful to pay attention as closely as you can.

- **Distractions.** Anything from too much noise and activity on the site to problems at home can steal your attention. If the problem is loud noise, tell the speaker you can't hear and to please speak louder, or move away from the source of the noise. If you are preoccupied with personal problems, concentrate even more on what the person is saying. (It often helps to get your mind off those problems anyway.)

- **Your ego.** Do you finish people's sentences for them? Do you interrupt others a lot? Do you think about the things you are going to say instead of listening to what you're being told? That's your ego putting itself squarely between you and effective listening. Be aware of your ego and try to tone it down a bit so you can get the information you need.

Remember the lessons you've learned about respecting your boss and showing consideration toward your co-workers. Listening as closely as you can to them, no matter what the circumstances, is part of keeping up good relationships with the people you work for and with.

Listening in the Classroom

As you take this class in critical skills, you can start applying these listening skills. Is someone on the other side of the room speaking too softly? Then ask your classmate to speak a bit louder, or to face the class while talking. Is there noise out in the hall? Ask permission to close the door. Did your instructor say something you did not understand? Ask your instructor to explain it further.

Becoming an effective speaker

Effective speaking is more than just talking so everyone can hear you; it's making sure that everyone understands what you've said. You have to think about what you will say (the content), who you will say it to (the audience), and how you will say it (the presentation). When you are an effective speaker, your listeners will never have to wonder what you are talking about, whether you are talking to someone in person or speaking over the phone or two-way radio.

Speaking face-to-face

Most conversations you have on the worksite will be between you and your boss or individual members of your team. By being clear in what you say and considerate of the person listening to you, you can save yourself a lot of time and frustration as you are speaking.

Give clear instructions and directions. Make sure they aren't vague or open-ended. Suppose you are a plumber's apprentice and a more experienced co-worker says, "Hey, go and get me that, uh... the right tee for this fixture. You know what I mean, right? It goes over here." Where are you supposed to go? What are you supposed to get? On the other hand, if your co-worker gives you clear, detailed instructions—"Hey, go out to the truck

> "The trouble with talking too fast is you may say something you haven't thought of yet."
>
> – Ann Landers

and get me a 3" tee, and make sure you get one that's got a 1½"-inch inlet on the side"—you'll know exactly what to do. How many times have you heard street directions like, "Go down to the store and then go a little bit after that and turn down the street." Which store? How far is "a little bit after that"? What's the name of the street, and do you make a right or left turn? It's better to say, "Go past the Safeway, and then make a right onto Park Street, which will be at the second traffic light after you pass the Safeway. We're at 111 Park Street."

Consider your listener. When talking shop with an experienced co-worker, especially one you've worked with for a while, you can take some shortcuts, like using lingo, when you're speaking. However, you must speak more carefully to inexperienced workers or to workers for whom English is a second language. In these cases:

- Explain terms they may not understand.
- Speak a little more slowly and don't slur your words.
- Encourage questions.
- Demonstrate what needs to be done.
- Ask your co-worker to paraphrase back to you what you've said.
- Make a sketch if you think it would help the other person to understand.

Check your tone of voice. How you say something is often as important as what you say, so listen to yourself. Do you sound calm or angry, confident or uncertain, patient or impatient, upbeat or negative? When you are calm, confident, patient, and upbeat, your listeners will pay closer attention and respect what you say. They won't get uncomfortable or irritated. Which of the following speakers would you rather have teaching you?

- **Speaker A:** "OK, listen up. I don't want to have to go over this again. I'm a construction worker, not a teacher, see? So pay attention, otherwise you'll end up in the hospital. We're gonna use this thing here… first you do this… OK, yeah. That seems about right. Now, to set this thing up, you've got to… hmmm… OK, now this looks right. See what I did there? If you didn't, you better pay more attention. This busts up concrete which means it'll bust you up if you don't know how to use it."

- **Speaker B:** "I'm going to show you how to set up and use this pavement breaker. This tool can be dangerous, but I'll show you how to stay safe. So pay attention, folks. First, I'll go over the steps on how to use it; I'll do that twice, more if you need it. If you have any questions or don't understand something, ask! That's what I'm here for. Next, I'll turn the unit on and use it so you can see how it works and how to operate it. Then the part you are all waiting for: each of you gets your chance to practice using it. Before we get started, let's check our safety gear. You always do that first."

Speaking on the phone or two-way radio

When you communicate by phone or radio, you don't have the advantage of looking at the person while you are talking. You'll have to make a greater effort to ensure that you send and receive the right information. Here are some things to keep in mind when making or taking a phone call.

Making calls

When making phone calls, you will sometimes reach an assistant or voice mail. So before you make your call, come up with a brief message you can leave in case you don't reach the actual person. Here are some tips to follow when making a phone call:

- Give your name and your company name and say why you are calling; speak clearly and spell out your name and company name if necessary.

- If you reach the person you intended to, take notes (if it's practical) during the call to ensure that you get the right information.

- If you leave a message, keep it short, and make sure the message includes your name, your company's name, a contact number, and the best time to reach you. When you are leaving a message on voice mail, take care to speak slowly, especially when relaying such information as contact phone numbers. You may want to have this information written down before you make the call, so that you don't have to look for it while you are leaving the message.

Receiving calls

Etiquette is the proper or accepted way of doing something, and using the phone has its own etiquette. Keep in mind that some people don't communicate on the phone as well as others, so if callers don't identify themselves completely or say why they are calling, ask them to. Here are some rules of etiquette to follow when taking a phone call:

- Don't answer by simply saying "hello." Give your name and the company's name.

- Don't leave callers on hold. If you don't have time to talk, say so pleasantly and suggest a better time to call back.

- If you're on a phone that allows calls to be transferred, let the caller know you're transferring the call. Be courteous, and give the caller the number you are transferring the call to in case the call is disconnected during the process.

Using electronic communication devices

Electronic communication devices, such as cell phones, are everywhere these days. They can be a big help on a construction site, but there is a time and place to use them as well as situations when you shouldn't.

- Don't make cell phone calls or send text messages while driving or operating heavy equipment. Many states have made it illegal to use a cell phone while driving, since this has been shown to be a factor in accidents. The best thing to do is turn off your cell phone while driving. If you must make a call, pull over to a safe spot on the side of the road.

- Check with your supervisor or human resources department on your company's policies regarding the use of cell phones and other electronic communication devices on the job. Companies may limit employees' use of cell phones on the job or at a particular site for safety or other reasons.

- If you're in a meeting with people, especially if you are talking to your boss, it's usually considered impolite to take a cell phone call. In such cases, put your phone on the vibrate option so that the noise doesn't interrupt the meeting. If you must take a call when you are talking to the boss or customers, excuse yourself and move away.

- Keep your cell phone calls short and limited to business while on the job, even if it's your personal phone and not the company's. Cell phone companies have different ways of charging their customers, depending on the plan; some charge by the minute, others for both incoming and outgoing calls. If you have a company-provided phone or device, you must limit your use of that device as much as possible, and never use it for personal reasons.

- Make sure that any cell phone you use has been charged recently; it can be frustrating to be talking to someone and have the phone suddenly lose power.

Nonverbal communication and body language

Research has shown that up to 90 percent of all in-person communication may not be through words at all; it comes from what you are doing with your body while you are talking. *Table 1* lists a number of different types of body language and the interpretations people commonly apply to them. You may think you're saying all the right things, but your boss could be getting an entirely different message from what you're doing with your arms, hands, feet, head, or eyes. So be aware of your body language while you are talking to others, especially when speaking to your supervisor:

> "There's language in (the) eye, (the) cheek, (the) lip... at every joint and motive of (the) body."
>
> – William Shakespeare

Mean what you say. If you're sincere or are genuinely interested in what your boss has to say, chances are your body language will reflect that.

Stand up straight. Slouching makes you look like you are not listening or you are bored. Even if you are tired at the end of the day, stand up straight.

Avoid crossing your arms. When you cross your arms, you are sending several negative signals: "I'm tuning you out" or "I don't agree with anything you say" or "I don't like you."

Maintain eye contact. Eye contact indicates to the other person that you're involved in the conversation. When you don't maintain eye contact, it may be taken the wrong way. Even if you are being reprimanded, look your boss straight in the eye.

Avoid frowning or clenching your fists. When you do this, you may be sending a message that you are angry and defensive. That's liable to make the other person angry or defensive in response, which means the conversation won't be productive at all.

Table 6-1. Interpreting Body Language

Body Language	Interpretation
Standing with hands on hips	Readiness, aggression
Arms crossed on chest	Defensiveness
Rubbing hands	Anticipation
Tapping feet or drumming fingers, pen, or pencil	Impatience
Patting or playing with hair	Lack of self-confidence or insecurity
Tilted head, leaning forward	Interest
Stroking chin	Thoughtfulness, deciding
Biting nails	Insecurity, nervousness
Rubbing eyes	Doubt, disbelief
Sitting with hands clasped behind head, legs crossed	Confidence, superiority
Brisk, erect walk	Self-confidence
Turning your back to another person	Dismissal
Looking out of the corner of your eyes	Disdain, scorn

Summary

Effective listening and speaking clearly and concisely are critical to your everyday performance as a construction professional. As a listener, you need to stay focused on the speaker, keep an open mind about what's being said, pay close attention (and keep notes if possible), and ask for clarification as necessary. Don't let emotions, boredom, or distractions get in the way while you're listening. As a speaker, whether you are talking to someone in person or over the phone or radio, be clear in what you say and be considerate of your listener. Know how to make calls and leave voice mail messages.

Here's a quick quiz that allows you to apply what you've learned in this module. Select the best possible answer, given what you've learned.

1. Effective two-way communication requires _____.

    a. the speaker to deliver messages that are clear, detailed, and to the point

    b. the listener to avoid being bored, distracted, or emotional while listening

    c. the sender and receiver to have good listening and speaking skills

    d. All of the above

2. When you are listening to instructions, the best way to make sure you get all the information you need is to _____.

    a. take notes and compare them with co-workers' notes

    b. read a book about the topic later on

    c. ask questions, but only after the person has stopped talking, because it's rude to interrupt

    d. take notes if you can, ask questions at appropriate times, and paraphrase instructions back to the speaker

3. You are on a noisy worksite when a supplier phones you with questions about your supply list. Some quantities have been left out, and some items you wanted will have to be substituted. To make sure you get what you need, you should _____.

    a. talk louder or yell if necessary

    b. tell the supplier to call back in a few minutes so that you can find a quieter place to continue the conversation

    c. move to a quieter spot and estimate what you need, figuring you can always return any excess

    d. call the supplier back from a quieter spot with a copy of the supply list in hand

4. While demonstrating a new procedure, your supervisor notices that you're slouched against a tree and looking off into space. Your supervisor is likely thinking that you _____.

    a. are tired after a long day

    b. may have hurt your back

    c. must have seen something unsafe in the distance

    d. don't seem to care much about your job because you're not paying attention

5. You're new on the job, and you want to make a good impression. Your supervisor tells you to get a pipe clamp, but you have no idea what that is. You should _____.

    a. ask your supervisor to describe what the pipe clamp looks like and to suggest where you might find one

    b. look around the jobsite for a pipe-shaped thing with some kind of clamp on it

    c. ask someone else, preferably a more experienced worker, to get a pipe clamp for you

    d. look up the term *pipe clamp* in a construction dictionary

6. A more experienced worker is training you and a co-worker in the use of a portable cement mixer. Your trainer has an oversized ego and is boring to listen to. You should _____.

    a. respect your trainer's experience and knowledge, even if you don't like the presentation, and force yourself to pay attention

    b. tune out and ask your co-worker later for a quick recap of what the trainer said

    c. ask the trainer so many questions that you become irritating, and then have a good laugh about it with your co-worker later

    d. pretend you are paying attention and later go read the cement mixer's user's manual, because your reading skills are better than your listening skills anyway

7. You've been working with the same group of people for about five years. You are all friends, and you all know each other's working styles. Today it's your job to explain a new procedure to them. You can assume that your co-workers _____.

    a. will consider it a good opportunity to have a few laughs and will keep interrupting you to make jokes

    b. won't need an explanation of any technical terms you use

    c. probably won't listen very closely to you because they only listen closely to instructions when they're from the boss

    d. will resent you for trying to teach them something new

8. You're driving the company truck to pick up a load of materials. Before you leave, you are told to keep your phone on because the boss might have to give you additional instructions en route. As you are merging onto a busy freeway, the cell phone rings. You _____.

    a. answer the phone because you know it's work-related

    b. answer the phone by yelling, "I'm busy merging now, I'll call back later"

    c. turn the phone off

    d. let the phone ring, pull over to the side of the highway when it's safe to do so, and call the number back

9. Your boss tells you to go pick up some materials of various sizes and shapes at the materials depot, and to do it quickly. The boss hurriedly writes something down on a piece of paper and hands it to you, and you leave without looking at it. Once you get to the depot to obtain the materials, you realize you can't read your boss's handwriting. You don't have a cell phone. What should you do?

    a. Get together with the employee at the depot whenever that person has a free moment and together try to make out your boss's handwriting.

    b. Go back to the worksite empty-handed and if your boss asks why, hand the paper over and say, "How am I supposed to read that?"

    c. Ask the depot employee politely if you can use a phone to call the boss and ask for clarification on what needs to be picked up.

    d. Kick yourself over the fact you didn't look at the paper before you left, then make an educated guess about what materials you need and in what shapes and sizes.

10. Your boss asks you to call a subcontractor about mistakes made on a recent delivery. This subcontractor is often hard to reach, and chances are you'll have to leave a voice mail message. You want to make sure the subcontractor knows exactly what materials to deliver. What is your best course of action?

   a. If you reach the subcontractor's voice mail, leave a detailed but brief message that includes material details and a contact name and number for any additional questions.

   b. Leave a message asking the subcontractor to call you back so that you can discuss the problem.

   c. Leave a message telling the subcontractor to call your boss back ASAP, otherwise there'll be big trouble.

   d. Make the call and if you don't reach the subcontractor in person, don't sweat it because it's a management problem and not really your concern.

Individual Activities

Activity 1: Listening Actively and Asking Questions

In this exercise, you are dealing with a supervisor who does not give clear instructions and who often uses vague or confusing words or phrases. For each item, read the supervisor's instruction, and then put down the question (or questions) you'd ask to get the specific information you need to carry out that instruction. Then pretend you are the supervisor, and put down what you think a supervisor's response should be to those questions.

1. **Supervisor's instruction:** Adjust the thermostat.

   Your question(s): _____

   _____

   Supervisor's response: _____

   _____

2. **Supervisor's instruction:** Get the MSDS for that coating and make sure everyone reads it.

   Your question(s): _____

   _____

   Supervisor's response: _____

   _____

3. **Supervisor's instruction:** Tell a few of the workers to meet with me later this afternoon.

   Your question(s): _____

   _____

   Supervisor's response: _____

   _____

4. **Supervisor's instruction:** Anybody who comes to this job inebriated will be persona non grata upon recognition of such fact. We do not condone dipsomania at this concern and I have been authorized to terminate any such person PDQ.

   Your question(s): _____

   _____

   Supervisor's response: _____

   _____

Activity 2: Identifying Poor Listening Habits

The following are some signs of poor listening habits. Check any boxes that you think apply to you. Be honest in your self-assessment. With any boxes you check, create an action plan for how you can improve that listening habit.

Check All That Apply	Actions That Show Poor Listening Skills
❑	Making jokes when the other person is trying to be serious
❑	Arguing with everything another person says
❑	Interrupting others constantly
❑	Rolling your eyes when you doubt what someone says
❑	Thinking about other things while someone else is speaking
❑	Fiddling around with things or fidgeting when someone else is speaking
❑	Disagreeing constantly with another person's suggestions
❑	Finishing another person's sentences
❑	Changing the subject to something you're more interested in
❑	Looking at your watch while somebody is talking to you
❑	Jumping to conclusions before the other person is finished

My action plan for improving my listening skills:

1. _____

2. _____

3. _____

4. _____

5. _____

Activity 3: Understanding How Communication Works

The goal of this activity is to help you see how two-way communication works in real-life situations. Read the three situations and then, for each one, identify the senders, receivers, and the communication channels used. In addition, identify whether any noise—that is, an obstacle to communication—is present. For each situation, complete the grid that follows.

Sender(s)	Receiver(s)	Communication Channel(s)	Is noise present? If so, what is it?

Situation 1

At the construction site, there's a problem with drainage after a heavy rain. Since you've got a good reputation as a problem-solver, your boss asks you to take a look at where the water isn't draining and to take a couple of your co-workers with you. The boss gives you a quick but vague rundown on what may be causing the problem and how you might deal with it, but says that you can solve it more or less as you see fit as long as your solution doesn't take too long or cost anything. "Stay safe, too, and make sure your solution indemnifies the company," the boss says as you leave. You are a bit confused by what the boss said, but you like a challenge, and so you ask three co-workers for some help. The four of you walk over to where the water has collected. You tell the others what the boss said about the problem, and then you ask your team for suggestions. You notice one of your teammates seems more interested in watching an earthmover operate on the other side of the site than in focusing on the problem. Another teammate disagrees with everything you propose but doesn't come up with any alternatives, and a third teammate wonders why the four of your have been asked to solve this problem since together you have zero years of real engineering experience.

Situation 2

Your boss hands you a cell phone and says to you:

> **Boss:** Any moment now, you'll be getting a call on this phone from P.R., who's over at the Landoff site. When P.R. calls, drive on over there. They'll need you to pitch in, and you're also gonna deliver some supplies.

You have a lot to do, and you're not too pleased with this order, but you do as you're told. The boss starts walking away.

> **You:** Hey, boss, what kind of supplies?

> **Boss:** Piping, some tools...

Just then a bulldozer starts up nearby. You're pretty sure, however, the boss said, "P.R. will know." You leave the cell phone on, expecting the call. Time passes, and there's no call from P.R., and you're thinking they didn't need you or the delivery after all, and, a bit relieved, you continue with your duties. A few hours later, however, you get the call.

> **P.R.:** I'm over at Landoff...

> **You:** Sorry to interrupt, but I just realized I don't know where that is.

> **P.R.:** You don't? Ah, gee... well, you get on Seventh... I think it's Seventh Street... or maybe Avenue, doesn't matter, it's Seventh, go past the big highway and then turn and go a couple of miles until you see a big dig, can't miss it.

As P.R. is talking, you realize you have nothing to take notes with, so you go looking for paper and pencil. While looking, you ask P.R. to repeat the directions and to be a little more specific, but you can't understand P.R.'s response because your cell phone seems to be running out of power. You ask for a callback number, but the phone's gone dead. You can't find the boss to ask for the callback number, so you ask around the site until you find someone who knows P.R.'s cell phone number. After some 20 minutes, you've found a phone that works, a pen, and something to write on, and you call P.R. back.

P.R.: (*irked*) Where the heck are you? I thought you left 20 minutes ago, are you lost . . .

You: I still don't know how to get there or what kind of supplies...

P.R.: (*shouting*) You haven't left yet? What the... it's the Landoff site, piping and drills, the boss filled you in...

You: (*loudly*) The boss said you would... so what size pipe, what kind, what kind of drills, I can't just waltz out of here with 'em...

P.R.: (*yelling*) I don't know, they don't tell me anything.

You: Can you go find out?

P.R.: No, I can't go find out, you were supposed to be here half an hour ago.

You: (*yelling*) You just called a few minutes ago, how can I have been there a half-hour ago... can't you get the facts straight?

P.R.: (*angrily*) Can't you get yourself in gear for once? Get over here, NOW! (*Hangs up*)

You grab a couple of tools and a few odd pieces of pipe that are lying around, and head off in the direction of where you think the Landoff site is.

Situation 3

During a pressure test on a newly installed plumbing system, you detect a leak. You call your boss, who's solving a problem elsewhere on the site, on the two-way radio. You say, "Boss, there's a leak in the system." Your boss is standing near some trucks that are backing up to unload supplies. There is also a crew of laborers nearby who are breaking up an old driveway. Your boss asks you to repeat your message and to speak up so that you can be heard over the backup beeps and jackhammers. You repeat your message, and your boss replies, "Keep the water running, trap the leaking water, and ..." but then a jet leaving a nearby airport roars over your head, drowning out the last part of your boss's message. When you try to call back, your boss has turned off the radio. You figure the boss thinks that you got the entire message and can't be bothered with any more questions. But your boss doesn't know that the jet passing overhead drowned it out.

Activity 4: Listening and Following Directions

Materials Required

• Paper (at least two sheets per team) and pencil

• Two drawings (provided by your instructor)

In this activity, you will see how hard it can be to communicate well. You will also get a chance to practice your speaking and listening skills. Work in teams of two. One team member will be the communicator and the other will be the receiver. Once you complete this activity, switch roles with your partner so that you both get a chance at sending and receiving information.

To complete this activity, follow these steps:

Step 1 Sit back-to-back with your partner.

Step 2 Your instructor gives a copy of a finished drawing only to the communicator. The receiver is not allowed to see the drawing. The communicator is asked to describe the drawing to the receiver as accurately as possible.

Step 3 As the communicator describes the drawing, the receiver draws a picture based on that description. The receiver can ask for an instruction to be repeated only one time, and the receiver cannot ask any other questions.

Step 4 The receiver stops drawing when the instructor says time's up and turns the completed drawing over.

Step 5 Partners switch roles. The instructor then gives the person who is now playing the role of communicator an entirely new drawing, and the partners repeat Steps 1 through 4, except that the two classmates have switched roles and are using a new drawing.

Step 6 After the instructor tells the receiver to stop drawing, the partners then compare their drawings to the originals and discuss the following questions:

1. How well did the receivers reproduce their respective original drawings?

2. How could the communicators improve their speaking skills?

3. How could the receivers improve their listening skills?

Activity 5: Listening and Following Directions

This activity may be used instead of Activity 4.

Materials Required

• Two sheets of lined paper

• Ruler

• Pencil with an eraser

In this exercise, your instructor will read you two sets of directions. Each set of directions will be read twice. You must listen carefully and draw or write things based on the instructions. You are not allowed to ask any questions. When everyone is finished, your instructor will show you an example of how the drawings should look based on the instructions that were given, and you'll have the opportunity to compare the drawing you did with how the drawing should look.

Discussion Questions

After the class finishes their drawings, discuss these questions:

1. Which instructions were easy to understand?

2. Which instructions were hard to understand?

3. What could the instructor have done to make the instructions easier to follow?

4. Do you think this exercise would be easier if you were allowed to ask questions?

Directions: Set 1

1. Take out one sheet of lined paper. Very lightly, draw lines to divide your paper roughly into four map sections: north, south, east, and west.

2. In the center of the southeast section, print the name of the only month that has three letters.

3. Count down three lines from the top of the northeast section. Between lines three and four, in the center of the space, write the year you were born.

4. In the center of the southwest section, draw a circle with a diameter of about 1 inch. Draw a vertical diameter inside the circle.

5. Count down nine lines from the top of the northeast section. On the ninth line, centered horizontally in the section, write in words the sum of three plus two.

6. In the southwest section, locate the circle you drew. Place your pencil at a point on the top left side of the circle and draw a line upward and to the left until it meets the top line on the left-hand side of your paper.

7. Turn your paper upside down and locate the center of the page. Draw a 1-inch square in the center of the page. Next, draw a triangle centered inside the square. No line of the triangle should touch any line of the square.

8. Turn your paper right side up. In the northeast section, locate the line where you wrote the sum of three plus two. Count down four lines. In the center of the line, print your last name.

9. Locate the line where you wrote the year you were born. Move your pencil over to the far left-hand side of the page. Draw a 2-inch arrow pointing northeast.

10. In the northwest section, count down six lines. At about ½ inch from the left-hand edge of your paper, print the 5th and 10th letters of the alphabet in reverse order. Place a period after each letter.

Directions: Set 2

1. Take out another sheet of lined paper. Very lightly, draw a line to divide your paper into a right-hand section and a left-hand section.

2. Locate the second line from the bottom on the right-hand side of the page. Print your full name preceded by the last four numbers of your Social Security number.

3. Count down three lines from the top of your paper on the left-hand side. Print your complete mailing address.

4. Locate the line where you wrote the last four numbers of your Social Security number. On the second line directly above that number, print the name of any high school.

5. Locate the lines where you wrote your address. To the right of your address, write your present age in months to the nearest month. Write your month age in numbers.

6. Did you attend the high school you wrote down earlier? Print your answer four lines above the last four numbers of your Social Security number.

7. In the center of the page, use your ruler to draw a rectangle that measures 2 inches vertically and 4 inches horizontally.

8. Inside the rectangle, on the left-hand side, draw a circle. The circle should touch the rectangle's left-hand side and the top and bottom of the rectangle.

9. Inside the circle, draw two curved arrows. Draw one arrow on the right side of the circle and draw one on the left side. The arrows should show a clockwise direction.

10. Count up three lines from the bottom of the page on the far left-hand side. Print a number and words to describe the area of the rectangle you drew.

Activity 6: Role-Playing Exercise—Leaving Phone Messages

No one likes to play phone tag. That's when two people can't seem to hook up with each other to have a phone conversation, so they end up leaving voice mail messages. If the messages aren't clear or detailed, they have to keep calling each other back for clarification. Phone tag can waste valuable time during the workday, so it's important to limit the number of times you have to call people and the number of times they have to call you. In this exercise, you'll split into groups of four, and each member of the group will read one of the following scripts, which represent four different types of voice mail messages.

Team Members and the Script Number They Read

1. _____   2. _____

3. _____   4. _____

Discussion Questions

After each classmate finishes reading his or her script, discuss these questions:

1. What did the caller do correctly?

2. What did the caller do incorrectly?

3. What tips for improvement would you give this caller?

Script 1

Hey T.J., what up. Look, we need those permits, you know, like pronto. Anyway, I got a spare ticket for the game this weekend, someone broke their leg, bummer, but you can't get up them stadium stairs with a busted flipper. These are sweet seats, but you gotta let me know by tomorrow so give me a call... oh, yeah, and about the permits—I almost forgot—we need them by Monday. And you gotta bring 'em over here, OK? So don't forget.

Script 2

Hi T.J., this is T.L. from Rocket Construction. We need the plumbing permits for the Basin Street project, and we've got to have them by 11 a.m. Monday morning. So can you drop them off beforehand? Our office is at 404 Mercants Avenue. That's M-e-r-c-a-n-t-s. To get here, take the Highway 20 freeway to the Sanderson Boulevard exit; that's the stadium exit. Go north on Sanderson, you'll pass the stadium on the left and then come to the intersection with Mercants Ave. You'll make a right on Mercants, and our office will be a little over two miles on the right. Look for the big sign with a rocket on it, that's us. Call if you've got questions. I'm at [*said slowly*] 703-555-1212. Have a good weekend, and thanks.

Script 3

T.J.! You are never there, dawg. What are you, allergic to the phone? I know you're not that busy. Yo, you know who this is, right? The boss is ragging on me about those permits, so I need them Monday a.m. or I'm cooked. You know, Basin Street. Monday, early, do I have to draw a map for you? If so, call. Later.

Script 4

T.L. here. Look, I'll make this short 'cause my cell phone bill is through the roof, you wouldn't believe what it was last month. Anything over 100 minutes they gouge you, man. And then if you call on the weekend, that's when they really nail you. I shoulda gone into telecommunications, like my uncle did, he's doin' very well, I mean he's well past bein' rich, lives out past Basin Street... oh, yeah, that reminds me. Plumbing permits. Monday morning, We're out by the stadium, you know, on Merchants? That's with an "h" I think... heck, just look for the sign, everyone knows it. It's like a local landmark or something, with the rocket? And the little sparklers, they got these colored lights goin' day and night... anyway, we need those permits, dude, so don't hang me out to dry with my boss... gotta go, I'm on my cell and I see a cop in my rearview...

Activity 7: Being Specific and Getting to the Point

We often fail to be specific when speaking to others. We assume they'll know exactly what we mean. Sometimes they do, especially if they know us well, but sometimes they don't. Read the following seven statements and fill in the blanks by writing down what you think are the clarifying answers. Next, review the three situations and then rewrite the statements so that they are clearer, using the questions following each statement as guidance. Finally, break into groups of three or four, and compare your answers to the statements and situations to those of your classmates. Did everyone provide the same clarifications?

Team Members

1. _____   2. _____

3. _____   4. _____

Statements

1. A supplier phones to say you will get your siding at the *end of the week*. This means that your siding will arrive _____.

2. Your boss says to turn in your time sheet at the *end of the day*. This means that you should turn your time sheet in by _____.

3. Your boss tells you to come to the office *first thing Monday morning*. This means you should go to the office at _____.

4. A subcontractor leaves this message on your voice mail: "I need those hazmat disposal permits *ASAP*!" This means that you must get the permits to the subcontractor on or by _____.

5. Your boss walks through the worksite and tells you to clean up an area *when you can get to it*. When should you clean up the area? _____

6. You tell your boss that a wall should be finished in a *little while*. When should your boss come by to see the finished wall? _____

7. A co-worker needs a ride and arranges to meet you *around 7:00 over by the ballpark*. Is this enough information for you to know where to pick up your co-worker? _____

Situation 1

You're standing next to your supervisor and you overhear a co-worker say the following:

Hey, boss? A moment, if you've got it? You know I've been working real hard and putting in a lot of OT lately, trying to meet the schedule. Not that I'm complaining, understand, this is a good job, I'm glad to have it. But it's been four consecutive weekends I've been on OT, you know? Anyway, if I can have a moment… the point is, my sister is getting married in a little while. You talk about one excited girl. Up in the Adirondacks, way up state; it's going to be this big three- or four-day affair, it takes a day just to get to the place. The guy she's marrying is pretty rich, I suppose; I think his family is in construction. The DiMimoso's? You heard of them? Well, he's a DiMimoso. Anyway, you know our daddy passed on a while back, or maybe you didn't know… long story short, she wants me to give her away at the wedding. I mean, what an honor, huh? Seeing that my daddy's not around to do it, you understand. So, I mean, is that OK?

Based only on what this worker has said, do you know what the worker wants? What do you suppose the worker is specifically asking the boss for? Do you think the boss has enough information to be able to make a decision?

Situation 2

You are taking a break with some co-workers when one of them says:

There's stuff going down, I've kept my eyes open, I don't know, but… I mean, just now, in with the boss, I say "Look here boss, you know the crew would go to the wall for you and all…" I mean, I just made a reasonable request, we've been kickin' it 10 hours a day or more for weeks it seems like. But the boss? Heh, fuhgetabout it, I mean, if looks could kill, I'd be on a coroner's slab now. "One more word," the boss says. I mean, what up with that, huh? But I didn't trot out of there like a trained puppy, either, just because it got a little hot. Mark my words, there's more here than meets the eye. It's like the X-Files. So, you all with me on this?

Based on what your co-worker has said, what are you supposed to do? What do you think the co-worker is trying to say to the group?

Situation 3

A co-worker tells you not to inhale a solvent, and you ask why. The co-worker replies, "Go ahead and try it, you'll see." Based only on what your co-worker has said, do you know what actually will happen if you inhale the solvent?

Rewrite

When you have finished the discussion, rewrite each statement so that someone hearing it knows exactly what is meant.

Activity 8: Finding Ways to Reduce Communication Noise

In this activity, you will come up with an action plan for dealing with communication noise. Choose one or more of the following situations, then work with three or four classmates to develop an action plan.

Team Members

1. _____   2. _____

3. _____   4. _____

Situation 1

You've been using your credit cards a bit too freely recently, and now you're in a lot of credit card debt. Also, you lent money to a relative, and then your car needed some expensive repair work. Your money worries are keeping you awake at night, and you are coming into work tired and unable to focus. Yesterday, you did not hear some instructions your boss gave you and as a result made several mistakes. You need a plan to deal with your money problems and concentrate better at work.

Situation 2

You get excellent grades in your classroom work and have been told you are the best trainee at the company. Now you're being asked to tutor other trainees. So now you think that you don't have to listen as much as the other trainees do. Your boss says for all trainees to go to a safety refresher session, but you figure you know all that already or can pick it up on the job quick enough, so you skip it and miss out on some important updates. As a result, you do something wrong at work and one of your co-workers is almost seriously injured. You need a plan to help you deal with your ego getting in the way of listening and learning.

Situation 3

Your boss has teamed you with an experienced co-worker to enhance your on-the-job training. This trainer seems nice enough until the training starts, then it goes downhill. Your trainer doesn't give you a chance to ask questions and loses patience easily. When you don't catch on to something immediately, the trainer is vaguely insulting. After a while, you just start tuning out, which makes the relationship even worse. Both you and the trainer go to the boss to complain. The boss tells you to work it out on your own, but that the training must continue. You need to do your part by making a plan to stay focused and not get so upset by your trainer, so that you can get the most out of the sessions.

Situation 4

Your boss is chewing you out because you've made a lot of mistakes lately and you left early three times this week. You left early for good reason, however—to go visit your father in the hospital, where he's recovering from open-heart surgery. Naturally, you're worried. Your boss starts giving you instructions in a rather nasty tone of voice on how to operate a piece of machinery. You are angry with your boss for being unfair, and you don't like it that some of your co-workers can hear the boss yelling at you. You're thinking about all the times you've worked well and showed up on time, but that doesn't seem to matter to your boss. Internally, you're griping at the boss instead of listening to the instructions. You need a plan for dealing with your emotions and for getting those instructions repeated without making your boss even angrier.

Activity 9: Pass It On

This activity is based on a traditional party game in which people pass along a story and as one person tells it to another, the story's details gradually change because everyone tells it a little differently. At the end of the game, the story is quite unlike the original. This activity is designed to show how hard it is to keep information accurate as it moves from one person to another. You can work in groups of four, or your instructor may decide to divide the class into two larger groups.

Instructions

1. One trainee should quietly read the information below about safety gear. The rest of the class must close their books.

2. This trainee should then close his or her book, and, in a voice low enough so that the other trainees can't hear, relate the information to another trainee, while keeping the book closed.

3. The trainee who hears the information should listen carefully and then, in a low voice, repeat what he or she heard to the next trainee, again while keeping the book closed.

4. The process repeats until it comes to the last trainee in the group, who repeats the information aloud to everyone in the group. Before the last person repeats the information, everyone should open their workbooks to check how the statements compare against the original.

Information for Activity 9

"To stay safe on the job, you have to wear appropriate personal protective equipment, including a hard hat, gloves, and eye and ear protection. If you don't wear the appropriate safety gear, you could get into trouble with your boss. Even worse, you could be injured or killed."

Discussion Questions

1. What surprised you about this activity?

2. What vital details changed during the telling of the information?

3. What could you have done to relate the information more accurately?

4. What could you have done to listen to and understand the information better?

Communication Skills II: Reading and Written Communication

"Outside of a dog, a book is man's best friend. Inside of a dog, it's too dark to read."

– Groucho Marx

Introduction

Look around your jobsite; notice how much written information there is. There are safety guidelines, maintenance and operating instructions for equipment, and work orders. There are materials lists and your employee manual. It's pretty evident, then, that to readily understand safety procedures, operating instructions, supply lists and work orders, manuals, and other documents, you'll need good reading skills.

You'll need some basic written-communication skills as well. This means knowing how to compose easy-to-understand messages and concise notes. Even filling out your timecard requires a level of written-communication skill. Because people are using email and text messaging more than ever to exchange information, it's important that you know how to use technological devices appropriately as tools for written communication. By having sound written-communication skills, you'll be able to take on more responsibility: to write up such things as supply lists or change orders, or to send notes or memos to subcontractors or clients. Say you have a suggestion for improving something at the jobsite; such ideas are often best presented in writing. If you want your suggestion to be taken seriously, you'll have to know how to write up your idea concisely so that it's easy to understand.

In this module, you will learn how to read more effectively and write more clearly by applying some sound and basic reading and written-communication habits.

Reading on the job

A lot of your job-related reading will be about safety. To make the workplace safer, Congress passed legislation in 1970 that created the Occupational Safety and Health Administration (OSHA). You'll hear a lot about OSHA regulations on the job; one of the first things you'll probably read will be the OSHA safety poster, which is displayed at every construction site. This poster outlines your rights and safety responsibilities on the jobsite, and it's a critical responsibility of yours to be able to understand what it says.

Here are some other documents you'll be reading at work:

- Work orders
- Work permits
- Change orders
- Supply lists
- Bills
- Codes
- Material safety data sheets (MSDSs)
- Operating instructions
- Blueprints
- Maintenance manuals
- Information about company rules, pay, benefits, and leave policies

To read effectively, it's important to practice some fundamental habits; by applying these, you'll be able to understand all kinds of printed material. Here are some good habits practiced by successful readers:

Learn how to find the information you need. Some documents may seem complicated, but they'll be much easier to understand if you can find the information quickly:

- Look for cues. Cues are the tools of the reading trade, and writers use these tools as aids for readers. Examples of cues include a table of contents; the use of **boldface**, *italics*, and underlining to stress important words or phrases; section headings or chapter titles; numbered and bulleted lists; illustrations; tables; references; and indexes. For instance, you can use your employee manual's table of contents to quickly find the information on company benefits and rules. This Workbook uses a lot of these cues. Can you find examples of them? (Hint: many of these cues are used on this page.)

- As you search a document for the information you need, ask yourself such questions as:

  - Does this part of the document apply to me and the work I am doing?

  - Is this part of the document relevant to doing my job safely and well?

  - Are these directions about something I'll need to use?

  If you answer "yes" to any of these questions, then you've found something relevant to your job, and you should read it carefully.

- If you come across words or phrases you don't understand, make note of them and ask your supervisor or a more experienced co-worker what they mean. If there's a construction dictionary handy and you have the time, you can look up the word; however, it's generally best and quickest to ask your supervisor, especially about any terms or words that might be related to safety.

More information on how to find what you are looking for in books and technical manuals is included in *Appendix A* of this module. Some advice on how to read plans and blueprints is included in *Appendix B*.

Take notes. As you read the document, write down the main points. This technique will help you concentrate and remember more of what you've read. Unless it's your own personal copy, don't mark up the book you're reading by underlining important text; other people will be reading the same book after you're finished with it, and they don't want to have to wade through your notes.

Adjust your speed. Don't try to read too quickly; do it at a speed that allows you to absorb what you are reading. If you don't understand something, try reading it again—even if you're in a hurry. It's important to understand such things as operating instructions and safety procedures, and no one is going to give you a hard time if you take an extra minute or two to ensure you understand safety rules or operating regulations.

Visualize. Create a mental picture as you read. This is especially helpful when you are reading step-by-step instructions. Try to see yourself doing each step as you read it. You'll find the instructions easier to remember, and you'll be able to perform the task better.

Read all the directions first before starting. This technique is helpful when following step-by-step instructions. Don't try to perform a task by reading each step and doing what it says; read all the steps through first. Specific instructions make more sense when you have an idea of the big picture, which you get from reading all the directions through before acting on any individual instruction. This habit will save you a lot of time and trouble.

Ask for help when you need it. Some of the things you read won't be clear even if you read them a second time. If that's the case, ask your supervisor or a more experienced co-worker for help. The problem might not be your reading ability; the problem might be that what you're reading is poorly written!

Help yourself. Don't let a problem with reading hold you back. If it's a matter of words or phrases you don't understand, you can look them up in a number of different places:

- Use the dictionary or the internet.
- Use internet search engines such as Google.
- Use an online encyclopedia such as Wikipedia.

If it's reading in general that you find difficult, don't worry; you can take a number of steps to improve your skills. You can take a course in improving reading skills at a local community college or through a local adult education program. If you can't afford classes, check with your company's human resources department, if they have one; the company might offer courses on how to read better, or they might offer financial assistance to pay for courses taught elsewhere.

Study Tip

You may need to receive special training to use certain tools or perform certain tasks such as hazardous materials disposal. That training may require you to read textbooks or training manuals. Set aside some time each day to study these documents. Read in a quiet place, with no distractions. Take careful notes or highlight important information. By following these study tips, you'll get a better grasp of the material, and you'll do it more quickly.

Writing on the job

No matter what you do in the construction industry, you'll need some basic written-communication skills. When you're first starting out, you'll probably not have to write up anything more involved than a materials list or possibly a short note. As you advance in your construction career, you may be asked to compose longer, more complex documents, such as memos or reports.

To communicate effectively through writing, you don't have to be an author who knows a lot of fancy words and how to write complex sentences. You just have to develop some basic skills and then apply them. Whether you are writing a note to your boss, emailing a materials request to a supplier, or jotting down directions that you're faxing to another construction site, the idea is to get your point across clearly and quickly so that your readers will get your point with a minimal amount of effort.

Reading and writing go together; if you've mastered the basic reading skills discussed earlier, you already have a good idea on how to communicate in writing.

Here are some tips on how to write something that others can easily understand:

Give your readers the information they need. Recall the discussion of how helpful cues are when reading. To help the reader along, use these cues when writing something. For example, let's say you're about to start on a bathroom remodeling job and your boss tells you to quickly summarize for the customers the various options of paint and fixtures available. Here are two examples of how you could do it. The first example is well-written, but the second example is not only well-written, but makes the information easier for the customer to quickly grasp, by using cues such as bullet points, boldface type, and column headings.

Example 1

Here are the paint colors and faucets available for your bathroom: Color # 1415, Soft Jade; Color # 1416, Garden Moss; and Color #1417, Forest Glen. All are available in semi-gloss or eggshell finish. There are also three faucet sets: the Meridian (single handle) $109.88; the Mermaid (dual handles) $83.50; and the Monitor (dual handles) $95.75. All are available in polished brass or polished chrome. I've included paint samples and photos of the faucets. Please tell me your choices by Friday. If you have any questions, call me at 703-555-1212.

Example 2

Here are the paint colors and faucets available for your bathroom (paint samples and photos are enclosed). Please tell me your choices by Friday. If you have any questions, call me at 703-555-1212.

Paint Colors (Available in semi-gloss or eggshell finish)

1415 Soft Jade

1416 Garden Moss

1417 Forest Glen

Faucet Sets (Available in polished brass or polished chrome)

Model	Price	Handle Style
Meridian	$109.88	Single
Mermaid	83.50	Dual
Monitor	95.75	Dual

Define words or phrases readers might not understand. Even better, avoid using words that are hard to understand. If you are writing something for a co-worker, supervisor, or supplier, it's probably okay for you to use construction jargon. But it's not a good idea to use such terms with clients. Avoid trying to impress your reader by using unnecessarily big words.

Be concise. Make sure you get to the point in your note or message. Don't try to be funny; humor usually falls flat when you're writing about business. Be as brief as possible. Use these guidelines:

> "Brevity is the soul of wit."
>
> – William Shakespeare

- Tell the reader the reason for the message.

- If you're asking the reader to do something, make sure the reader knows what to do, when to do it, where to do it, and, if necessary, how to do it.

- Give the reader a contact name and number, in case there are any questions.

- Don't forget to be courteous. Remember you are a professional; even an informal note or message is no place to be sarcastic. Make sure you've written nothing offensive, even by accident—be particularly careful when you compose emails.

Visualize. If you can see something in your head, you can describe it to someone else. Let's say you have to write down and fax driving instructions to a supplier on the other side of town. This is how you could do it:

- Imagine how you'd drive the route.

- Write down the directions, including all the turns in order. Use specifics such as street names or route names instead of generalities such as "make a left when you come to the red house."

- Read over your directions and mentally drive the route again. Then ask yourself, "If I were from out of town, could I follow these directions?"

Check what you've written for accuracy before you send it. Whether it's a handwritten note, a supply list, or a letter, get into the habit of looking over what you've written. This is particularly important with email, which is so easy to send by clicking a mouse; read your emails carefully and use spell check before you send them out. Make sure you've identified yourself and the reason for the message or note, and make sure the reader will know what to do in response. If there's any math, make sure it's correct. Make sure you are consistent in the use of such things as product numbers or prices; and make sure you haven't left anything vital out.

Sometimes it is helpful to ask a co-worker to review something you have written, if that person has the time. Another reader can usually spot any mistakes or problems you might have overlooked. Don't be embarrassed to ask for this type of help, and don't be upset if your co-worker spots a mistake. (After all, professional writers employ editors to do this for them.) The idea is to catch mistakes in a document or message before it gets to the reader.

Summary

There are many important documents to read on a construction site, so you'll need to be able to absorb this information readily and thoroughly. Good reading habits include knowing how to quickly find the information you really need, taking notes, visualizing instructions and reading them through first before executing them, and asking for help with unfamiliar terms or anything you don't understand. You don't have to be a great writer on a construction job, but you should have some basic written-communication skills, even if it's just for filling out forms. By knowing how to present information concisely so that it's easy to understand, and by checking what you've written, you'll get your point across in your note, memo, or email.

Here's a quick quiz that allows you to apply what you've learned in this module. Select the best possible answer.

1. It's your first day on the job, and you've been given a copy of the employee manual. You're interested in finding information on the company's leave policy. To find the information you need quickly, it is best to _____.

    a. flip through the manual until you spot the words *leave policy*

    b. read the entire manual in a quiet location until you find the section on leave

    c. look in the table of contents to find the page where leave policies are described, or look in the index for words such as *leave* or *vacation* or *sick days*

    d. ask your boss or an experienced co-worker about the company's leave policies

2. Your boss tells you to read the manufacturer's safety instructions for operating a reciprocating saw. The safety instructions include words you don't understand. To operate the saw safely, what is your best course of action?

    a. Look up the words on an online construction dictionary.

    b. Figure that if you don't understand the words, the instructions probably don't apply to you.

    c. Ask your supervisor for help.

    d. Tell your supervisor that the instructions are difficult to understand, and that's the reason why the company should pay for reading classes for its employees.

3. You're looking through a large manual for information on how to operate a sheet metal brake. The manual includes step-by-step instructions for safely operating the brake. What is the best way to remember the instructions?

    a. Read the instructions more than once if necessary.

    b. Read the instructions and visualize yourself doing the tasks.

    c. If possible, read in a quiet spot that is free from distractions.

    d. All of the above.

4. Your boss authorizes you to set up a scaffold. The manufacturer's guidelines include setup instructions, safety instructions, and tagging instructions. What is your best course of action?

    a. Read only the setup instructions, because that's all you're doing—setting it up—and reading anything else will delay getting the job done.

    b. Perform each instruction in order as you read, double-checking what you've read as you go along.

    c. Read both the setup and safety instructions before putting up the scaffold, and after the scaffold is set up, go back and read the tagging instructions.

    d. Read all of the instructions, get answers to any questions you have, and determine what tools you'll need, before setting up the scaffold.

5. You should always carefully review your emails before sending them because _____.

    a. you can't use cues with emails, and so you need to triple check your content

    b. emails are expensive to send, and you don't want to have to send them more than once

    c. the spell checker is completely unreliable

    d. there is no Undo on the send button

6. You have to write a memo to get answers to questions about a task and to request supplies. To write a memo the reader will readily understand, you should _____.

    a. say who you are, why you are writing, what you are writing about, when you need an answer, and where and when you need the supplies

    b. arrange your questions in columns with bold headers, and ask a co-worker for more ideas on how to make the memo look cool

    c. introduce yourself, ask your questions, give complete delivery information, and then conclude with a joke or two about construction workers

    d. describe the project at some length in your memo, to provide helpful background for the supplier to answer your questions

7. You have decided to go into business for yourself as a painting contractor, and you want to come up with the best way to show clients what services your company offers and how much those services cost. The most effective method to present this information would be to create a(n) _____.

    a. advertisement with high-resolution photographs showing painters at work on beautiful projects

    b. memo with your company's name and contact information, with a table listing services in one column and prices in another

    c. memo that includes the date, your name, and a description of your company including an example or two of what you can do

    d. formal letter introducing yourself and describing your company and its services

8. If you are writing to a co-worker or supplier, it is okay to use construction jargon.

    a. True

    b. False

9. Your boss tells you to send an email with directions to a construction site in a nearby town to several suppliers who are all located in the same part of your city. What is the best way to complete your task?

    a. Call the suppliers instead, since you're better at giving directions when talking rather than in writing.

    b. Visualize the route starting from where the suppliers are, and then write down the directions from that part of town to the construction site, using specifics like street names. Then, consult a map to make sure you have the names and turns right before sending the email.

    c. Draw a map with the route the suppliers should take, scan the map into the computer, and then send out the email to the suppliers with the map as an attachment.

    d. Send a brief email that lists the streets to take, and include your supervisor's cell phone number in case anyone gets lost.

10. Your boss tells you to contact a subcontractor about a delivery that is overdue, reminding them that your company is still waiting for the promised supplies. Your boss tells you to fax the letter rather than send an email, because a fax will probably get the subcontractor's attention more quickly. How should you complete this task?

   a. Write a note asking the subcontractor somewhat sarcastically whether they've lost the directions to the site again, and whether they need one more reminder on how to get there.

   b. Send a fax with an illustration of a target, with the subcontractor company's name in the middle.

   c. Write a note telling the subcontractor that they'll face legal action unless they deliver the supplies requested, then ask a co-worker to read the letter over for any grammatical mistakes before you fax it.

   d. Write a note courteously reminding the subcontractor that your company is still waiting on the delivery. Obtain a copy of the original order, and restate the details in your note. Thank the subcontractor for their prompt attention to the matter, get your boss to sign the letter, and then fax it over.

Individual Activities

Activity 1: Understanding What You Read

This activity checks your ability to read two kinds of documents you'll come across regularly in your job: operating instructions and company messages to its employees. The first reading exercise is about the safe use of fire extinguishers; the second is a memo about an employee meeting. Read each closely and carefully, and answer the questions that follow.

Reading Exercise 1

Fire Extinguishers

Four classes of fuels can be involved in fires. Each fuel class requires a different kind of fire extinguisher. The four classes are:

- *Class A fires: Paper, wood, or other common combustibles*
  Class A fire extinguishers are filled with water. Because water can cause other types of fire to spread, you must never use Class A extinguishers on any other class of fire.

- *Class B fires: Liquids and gases*
  Class B fire extinguishers are filled with carbon dioxide (CO_2). The CO_2 smothers the fire.

- *Class C fires: Electrical equipment*
  Because electrical currents are so dangerous, Class C extinguishers are designed to protect against possible electric shock. Like Class B extinguishers, Class C extinguishers put out fires by smothering them.

- *Class D fires: Metals that burn*
  Class D extinguishers contain a special powder that prevents oxygen from reaching the fire. Because fires need oxygen to continue burning, the lack of oxygen will put the fire out.

Using a Fire Extinguisher

There are eight simple steps to using a fire extinguisher properly. You must always match the type of fire extinguisher to the type of fire.

Step 1 Hold the extinguisher upright.

Step 2 Pull the pin. The plastic seal will break.

Step 3 Stand at least 10 feet from the fire. The extinguishing material will shoot from the nozzle with a certain amount of force. If you stand too close you may blow burning objects into the air, thus spreading the fire.

Step 4 Aim at the base of the fire.

Step 5 Squeeze the handle to discharge the contents. Sweep the extinguisher from side to side.

Step 6 When the fire is out, move closer to look for any remaining burning residue.

Step 7 When the fire is completely out, quickly check to make sure it has not restarted.

Step 8 Leave the area quickly. Fumes and smoke can be harmful, even deadly.

Reading Exercise 1 Questions

1. The article includes several features that make this material easier to read. The features include _____.

    a. italic type, boldface type, a glossary, and charts

    b. boldface type, italic type, bulleted lists, and numbered lists

    c. boldface type, italic type, underlined words, and numbered lists

    d. numbered lists, figures, a chart, and bulleted lists

2. An important point is repeated twice, in slightly different ways, in the article. The point is that _____.

    a. fires need oxygen to keep burning

    b. smoke and fumes can cause serious harm to people

    c. it is important to match the type fire extinguisher to the type of fire

    d. carbon dioxide can be used to put out Class B fires

3. Under the heading *Using a Fire Extinguisher*, you will find step-by-step instructions for fighting a fire. The best way to remember this material is to _____.

    a. practice the steps by doing what they tell you to, including actually using the fire extinguisher in Steps 2 through 5

    b. visualize the steps in your mind as you read them

    c. give the steps a quick read-through but don't spend much time at it, because there's no fire now

    d. read all the steps through while you're doing something else, because it's not that important for you to know how to use a fire extinguisher; fighting fires is the responsibility of the fire department

4. Which types of fires are put out by smothering?

    a. Class B and Class C

    b. Class A and Class B

    c. Class C and Class D

    d. Class A and Class D

5. According to the article, when using an extinguisher you should stand at least 10 feet from a fire because _____.

    a. you might be burned if you stand any closer

    b. you might be overcome by smoke if you stand any closer

    c. the fire extinguisher won't work if you stand closer

    d. the force that pushes out the foam can blow burning material into the air

---

MEMO

All State Construction Company
#350 King Street, New Orleans, LA
Phone: (504) 555-8980   Fax: (504) 555-6542
email: AllState@aol.com

To: Hourly Employees of All State Construction

From: Human Resources Dept.

Date: May 5, 2009

Subject: Employee Meeting

A meeting will take place on May 12, 2009, at 8:30 a.m. The location is 450 Dauphin Street, second floor, Room 203. All hourly employees are required to attend. Workers who need to remain on the jobsite for safety purposes must inform their managers that they will not be able to attend. There will be a make-up meeting for those employees, at the same address, at 4:30 p.m. on the same day.

We will discuss the following important points at the meeting:

1. The absenteeism policy. Absenteeism has been very high lately.

2. Safety policies, especially with respect to working in confined spaces. We have seen many violations of safety procedures.

3. Tardiness policy. Our records show that workers are routinely punching in 15 to 20 minutes late in the morning and are returning from lunch 10 minutes late.

The meeting will take approximately an hour. All employees must report back to the jobsite as soon as the meeting is over.

Reading Exercise 2 Questions

1. Who must attend the meeting?

2. What is the only acceptable reason for missing the meeting?

3. You have promised to call a contractor at 11:00 a.m. on May 12. Will you be able to do so without missing the meeting?

4. You are a salaried employee of All State Construction. That is, you are not paid by the hour. Do you have to attend the meeting?

5. What do you think would happen if you showed up to the meeting late?

Activity 2: Reading Tables

Writers often use tables to summarize important information or to make certain types of information easier to read. In this exercise, you will test your ability to read a table. Review the following table, and then answer the questions that follow.

Types of Wood and Their Properties				Holding Power	
Type	Color	Weight	Workability	Screws and Nails	Glue
Ash	Light brown	Heavy	Hard	Average	Average
Balsa	Light brown	Light	Easy	Poor	Good
Birch	Medium brown	Heavy	Hard	Good	Good
Cedar	Red	Light	Easy	Good	Good
Cherry	Light to dark red	Medium	Good	Excellent	Good
Fir	White to medium red	Medium	Good	Good	Average
Hemlock	Buff or tan	Light	Easy	Good	Good
Mahogany	Red to dark brown	Medium	Easy	Good	Good
Maple	Light brown or tan	Heavy	Fair	Average	Good
Oak	Tan to red	Heavy	Hard	Good	Average
Pine (Eastern)	Light brown	Light	Easy	Good	Good
Redwood	Reddish brown	Light	Good	Good	Good
Walnut	Deep dark brown	Heavy	Good	Good	Good

Table Questions

1. Of the woods listed, _____ has the best holding power for screws and nails and _____ has the worst holding power for screws and nails.

2. The five lightest woods are _____.

3. You want a wood that is red, light, and easy to work, so you choose _____.

4. You want a light-brown wood that can hold both glue and nails, so you choose _____.

5. True or false: Mahogany is easy to work with.

Activity 3: Reading a Project Manual

A project manual is a book produced for large construction projects. The manual includes detailed information about every part of the project, from the size and type of construction materials to the color and type of paint. The following is an excerpt from a project manual on the construction of a medical clinic. Read the excerpt carefully and answer the questions that follow.

Project Manual—Excerpt

Project:	**Banister-Lieblong Clinic of the Conway Regional Medical Center, Conway, Arkansas**
Architect:	**Williams and Dean Associated Architects**
Structural engineers:	**Professional Engineering**
Mechanical engineers:	**Mike Trusty Engineering**
Electrical engineers:	**Harp Consulting Engineers**

Section 16471

Telephone Service Entrance

Part 1   General

1.01   Summary

A. This section covers the work necessary to provide and install a complete raceway system for the building telephone system. The system shall include telephone service entrance raceway, equipment and terminal backboards, and empty ¾" conduits to telephone outlets.

1.02   Related Sections of the Project Manual

A. Section No.   Item

09900	Painting and Staining
16111	Conduit
16130	Boxes
16195	Electrical Identification

1.03   Quality Assurance

A. Telephone Utility Company: Southwestern Bell Telephone Company

B. Install work in accordance with Telephone Utility Company's rules and regulations.

Part 2   Products

2.01   Telephone Termination Backboard

A. Material: Plywood

B. Size: 4' × 8', ¾" thick

Part 3   Execution

3.01   Examination

   A. Verify that surfaces are ready to receive work.

   B. Verify that field measurements are as shown on the drawings.

   C. Beginning of installation means installer accepts existing conditions.

3.02   Installation

   A. Finish paint termination backboard with durable enamel under the provisions of Section 09900 prior to installation of telephone equipment. Color to match wall.

   B. Install termination backboards plumb, and attach securely at each corner. Install cabinet trim plumb.

   C. Install galvanized pull wire in each empty telephone conduit containing a bend of over 10 feet in length.

   D. Mark all backboards and cabinets with the legend "TELEPHONE" under the provisions of Section 16195, Electrical Identification.

END OF SECTION 16741—TELEPHONE SERVICE ENTRANCE

Source: Nabholz Construction Corporation, Conway, AR

Project Manual Questions

1. Look in Section _____ of the project manual to find information on electrical identification.
   a. 16130
   b. 16195
   c. 19651
   d. 16471

2. _____ utility company provides telephone service to the Banister Lieblong Clinic.
   a. Southwestern Bell
   b. Pacific Utility Company
   c. Conway Telephone, Inc.
   d. Telephone Service Entrance

3. _____ will *not* be part of the complete raceway system for the building's telephone system.
   a. Terminal backboards
   b. Conduits measuring ½" to telephone outlets
   c. Telephone service entrance raceway
   d. Equipment backboards

4. The rules and regulations of _____ should be followed when installing the telephone service entrance.
   a. Southwestern Bell
   b. The Conway Medical Center
   c. Harp Consulting Engineers
   d. Williams and Dean Associates

5. The telephone termination backboard should be constructed of _____.
   a. ¾"-thick plywood
   b. ½"-thick plywood
   c. ¾"-thick medium-density fiberboard
   d. 4' × 8' galvanized steel

6. _____ should be used to paint the termination backboard.
   a. Semi-gloss latex
   b. Durable enamel
   c. Flat latex
   d. Off-white

7. The backboards and cabinets must all be marked with the legend "_____."

   a. COMMUNICATIONS

   b. RACEWAY RELAY

   c. TELEPHONE

   d. Telephone Service Entrance

8. _____ is the name of the electrical engineering firm working on this project.

   a. Mike Trusty Engineering

   b. Professional Engineering

   c. Harp Consulting Engineers

   d. Southwestern Bell Electric

9. Galvanized pull wire is to be installed in each empty telephone conduit containing a bend of more than _____ in length.

   a. 10 inches

   b. 10 feet

   c. 10.1 feet

   d. 8 feet

10. To find more information on conduits, you would look in Section _____ of the project manual.

   a. 09900

   b. 16111

   c. 161111

   d. 16195

Activity 4: Completing Workplace Forms

At work, you will fill out your share of forms: time cards, work permits, safety tags for equipment, injury reports, and requests for equipment or to take leave. Forms may be used for many different purposes, but certain procedures are used to fill out any kind of form: reading and following instructions carefully, filling out the form completely, and signing and dating the form.

This activity will give you practice in filling out two forms common in construction work: the weekly time card and the safe work permit.

Weekly Time Card

Your pay (regular and overtime) as well as your leave and benefits are based on the number of hours you work each week. It is important to complete your time card carefully and accurately. *Figure 7-1* is an example of a weekly time card, which you will fill out by following these instructions.

Instructions for Completing a Time Card

1. Complete the top part of the form by printing your name and Friday's date.

2. Use numbers, not words, when putting down your hours.

3. Fill in each day based on the following information. On Monday you spent 10 hours welding; 8 of those hours were regular time, and 2 of those hours were overtime. On Tuesday you spent 6 hours of regular time welding, and on Wednesday you spent 3 hours of regular time welding. The job number for welding is 14. The cost code for welding is A335.

4. You didn't work on Thursday—it was your day off.

5. On Friday, Saturday, and Sunday, you welded for 8 hours each day. You worked 2 hours of overtime on both Friday and Saturday to help the framing crew meet a deadline. The job number for carpentry work is 22, and the cost code is C625.

6. Total your hours for the week in the far right column and check your math.

7. Sign your name.

8. Ask a classmate to check your time card, make sure your addition is correct, and sign the time card for you as your supervisor.

Figure 7-1. Weekly Time Card

Emp. Name:_____   Week Ending:_____

NABHOLZ CONSTRUCTION CORP.
LABOR DISTRIBUTION

Job Number	Cost Code	Monday		Tuesday		Wednesday		Thursday		Friday		Saturday		Sunday		Total	
		Reg	O/T	Reg	O/T	Reg	O/T	Reg	O/T	Reg	O/T	Reg	O/T	Reg	O/T	Reg	O/T

Employee Signature: _____   Supt. Signature: _____

Source: Nabholz Construction Corporation, Conway, AR

Safe Work Permit

Safe work permits are issued by a qualified person: for example, a company's safety representative or the local fire marshal. These permits are required for certain types of tasks on a construction site. Examples of tasks that require safe work permits include:

- Welding or cutting operations that require the use of torches.

- Tasks that involve the possibility of fire or explosion.

- Tasks that must be done in confined spaces such as pits or tanks.

Authorized personnel must review and sign the permit. The permit allows authorized personnel to identify hazards, take steps necessary to safeguard workers, and make sure that the proper safety equipment is available and in good working order. Permits cover only the people identified in the permit for specific tasks and for specific dates and times.

In this activity, you will practice reading and completing a safe work permit. *Figure 7-2* is the example you will work with. To complete this activity, use the following information:

A note on terminology. This permit involves tasks known as hot work and lockout/tagout. Here is what these terms mean:

Hot work – Maintenance or repair tasks done on or near electrically energized or flammable equipment. Examples of this equipment include machines that operate by electricity, and pipes that transport natural gas or petroleum.

Lockout/tagout – This is a two-step method that ensures the safety of workers performing routine maintenance or repairs. First, workers place approved locking devices on a piece of equipment so that it cannot be accidentally turned on. Next, they affix a tag to the unit, which states who authorized the lockout and why and how long the equipment will be locked.

Section 1

- The permit is issued for May 22, 2009, from 8:00 a.m. until 8:00 p.m.

- The work to be performed is the repair of an electrical service that has shorted out.

- The building number where the work will be done is 7C.

- The building supervisor is M.K. Gold.

- The specific equipment is Electrical Panel #3.

- The specific location of the work to be done is Floor 1, Room 106.

- The company performing the work is Electrical Concepts, Inc.

- The workers performing the work are P. White (entrant), J. Green (entrant), and T. Brown (attendant).

Sections 2–13

(2) Type of permit issued: hot work

(3) Protective equipment required: gloves and goggles

(4) Preparation required: lockout/tagout and work-area barricades

(5) General instructions: Workers must report hazards or accidents immediately and must stop work if any alarm sounds.

(6)–(11) Special precautions: Workers must remove all combustibles (items that can catch fire) from the area before they begin working. They must also cover all floor and wall openings and sprinkler heads and keep a water hose running.

(12) Tests required: None

(13) Other test results: None

Figure 7-2. Safe Work Permit

SAFE WORK PERMIT
(1) GENERAL INFORMATION

Safe Work Permits are valid only during the shift (8 or 12 hours) for which issued. A new permit is needed for each shift.

This permit is Issued for: _____ (date)	from _____	AM PM	to _____	AM PM

Work To Be Performed:		Work Order Number:

Building Number or Department Name:	Building or Department Supervisor:

Specific Equipment or Object of the Work:	Specific Location of the Work:

Name of Company Performing Work:

Names of Employees Performing Work:	1– 2– 3– 4– 5– 6–		Circle if appropriate: (Confined Space)	Entrant Attendant

Entrant Attendant (×6)

(2) TYPE PERMIT ISSUED	Yes	No	Initials
Confined Space Entry			
Hot Work			
Line/Equipment Opening			
High Work			
Close Proximity Work			
Radioactive Device			
Other (specify):			
Insulation Removal—Separate Permit Required (IH PROCEDURE 7)			
Excavation—Separate Permit Required (SAFETY PROCEDURE 16)			
Hot Tapping—Separate Permit Required (SAFETY PROCEDURE 4)			

(3) PROTECTIVE EQUIPMENT REQUIRED	Yes	No	Initials
Ear Protection			
Gloves (Specify Type:)			
Acid Suit			
Protective Coveralls			
Respirator (Specify Type:)			
Goggles			
Faceshield			
Lifeline and Body Harness			
Safety Net			
Other (list)			

(4) PREPARATIONS REQUIRED	Yes	No	Initials
Lockout/Tagout			
Drain/Empty Line/Equipment			
Vent/Purge Line/Equipment			
Valve Off/Isolate Line/Equipment			
Blind/Disconnect Line/Equipment			
Barricade Work Area			
Barricade Floors Below			
Other (specify)			

(5) GENERAL INSTRUCTIONS	Yes	No	Initials
Instructed on Sign In/Sign Out			
Instructed on Hazardous Materials Involved in Work			
Instructed on Fire Extinguisher Location/Operation			
Instructed on Shower/Eyewash Location/Operation			
Instructed on Sounding Alarm/Evacuations			
Instructed to Stop Work If Any Alarm Sounds			
Instructed to Report Hazards/Accidents Immediately			
Instructed on Permit Requirements			

(6) SPECIAL PRECAUTIONS—HOT WORK (SAFETY PROCEDURE 5)	Yes	No	Initials
Inertize Equipment			
Remove Combustibles in Area			
Cover Combustibles in Area			
Cover Floor/Wall Openings			
Cover Trenches/Sewer Drains			
Cover Conveyor Systems			
Cover Sprinkler Heads			
Cover Combustibles—Floor Below			
Wet Down Floor/Floor Below			
Water Hose Running			
Type ABC Extinguisher at Work Site (10 lb. minimum)			
Instructions on Fire Watch Duties			

(7) SPECIAL PRECAUTIONS—CONFINED SPACE ENTRY (SAFETY PROCEDURE 2)	Yes	No	Initials
Outside Fresh Air Supply Blower/Vacuum			
Bottom Outlet Valve Open Position			
12 Volt Lighting or Ground Fault Interrupter			
Ladder Extended to Bottom and Tied Off			
Supplied Air Respirator w/ Built-In Escape			
Confined Space Watch/Attendant			
CPR & Rescue Personnel Available/Notified			
Entrants/Attendants Identified Above			
2 SCBAs, 2 Lifelines w/ Harness at Job Location			
Communications Device to Summon Help			
Management Approval for Off-Hour Entries			
Safety Dept. Approval when Hazards Can Develop			

(8) SPECIAL PRECAUTIONS—LINE OPENING (SAFETY PROCEDURE 4)	Yes	No	Initials
Support/Hanger Conditions OK			
Opening Lines Under Pressure			
Opening Lines w/ Combustibles			
Opening Lines from Ladder			
Opening Common/Shared Lines			

(9) SPECIAL PRECAUTIONS–HIGH WORK (SAFETY PROCEDURE 23)	Yes	No	Initials
Instructed not to Walk on Tank Tops/Pipes			
Instructed not to Walk on Cable Trays			
Instructed not to Walk on Fiberglass/Plastic			
Instructed to Inspect Safety Harness/Lanyard/Lifeline			
Permission Given to Enter Roof			

(10) SPECIAL PRECAUTIONS—CLOSE PROXIMITY (SAFETY PROCEDURE 23 & 27)	Yes	No	Initials
Lifting/Hanging Objects Over Lines/Equipt.			
Lifting/Rigging w/i 15 ft of Electrical Line			
Moving Equipment w/i 10 ft. of 50 Kv Line			
Moving Equipment w/i 20 ft. of >50 Kv Line			
Other (specify)			

(11) SPECIAL PRECAUTIONS—RADIOACTIVE DEVICES (SAFETY PROCEDURE 22 & 24)	Yes	No	Initials
Procedure for Removing Radiation Device Reviewed			
Procedure for Industrial Radiography Reviewed			
Safe Perimeter Established/Barricaded/Signs Posted			
Restricted Area Cleared of Personnel			
Approval of Radiation Safety Officer Obtained			

(12) TESTS REQUIRED	Yes	No	Results	Initials
Test Equipment Calibrated				
Initial Oxygen Meter Test (must be >18.5% and <22%)				
Continuous Oxygen Meter Test (must be >18.5% and <22%)				
Initial Explosion Meter Test (must be zero)				
Continuous Explosion Meter Test (must be zero)				
Toxic Substances (must be zero)				
Other (specify)				

(13) OTHER TEST RESULTS (ie., after breaks)	Time	Result	Initials
Oxygen [ ] Explosion [ ] Toxic (list) [ ]			
Oxygen [ ] Explosion [ ] Toxic (list) [ ]			
Oxygen [ ] Explosion [ ] Toxic (list) [ ]			
Oxygen [ ] Explosion [ ] Toxic (list) [ ]			
Oxygen [ ] Explosion [ ] Toxic (list) [ ]			
Oxygen [ ] Explosion [ ] Toxic (list) [ ]			
Oxygen [ ] Explosion [ ] Toxic (list) [ ]			

(14) REMARKS

(15) APPROVAL SIGNATURES INDICATING JOB INSPECTION AND PERMIT REQUIREMENTS HAVE BEEN MET FOR WORK TO PROCEED

Radiation Safety Officer (If Required)	Building Manager (Off–Hour Confined Space)	Safety Department (Variance Request)

Responsible Person for Maintenance or Contractor: Print: _____ Sign: _____	Responsible Person for Building or Department: Sign: _____

(18) SIGNATURES VERIFYING WORK COMPLETED AND WORK SITE READY FOR USE

Responsible Person for Maintenance or Contractor: Print: _____ Sign: _____	Responsible Person for Building or Department: Print: _____ Sign: _____

PLANT EMERGENCY PHONE NUMBER IS

Source: BE & K Construction Company, Birmingham, AL

(Notice that the permit has six sections for special precautions, one for each type of permit issued. You must fill out only the section for the type of permit requested here.)

(14) Remarks: The job was completed in two hours.

(15) and (18) Approval signatures: For this permit, no signature from a radiation safety officer is required. Signatures are required from the building manager, the safety department representative, the person responsible for maintenance, and the person responsible for the building. (For this activity, you can pretend that you are these people. Print and sign their names. On an actual job, of course, only authorized personnel may sign the permit.)

Activity 5: Getting to the Point in Writing

In this activity, you'll read an example of an email you have sent your boss. Then, you will answer the questions that follow.

From: You

To: The Boss

Re: Warehouse on Industrial Way

Date: May 12, 2008

Boss, how are things? I know you are busy, but I hope you will soon read this email. I'm sending it to tell you about the new warehouse that we are putting up. We're swamped, but I think we can stay on schedule. As you know, our crew is scheduled to start working there next Monday. I only hope that everyone shows up. Some of 'em keep coming in later and later. But nothing for you to worry about (I think). This is going to be a big job, and I've already put in a lot of overtime. I've got some things under control and I think the schedule ought to be OK, and there's a few things worrying me as well. Mostly the permits, which I know we absolutely, positively cannot start without. Anyway, hope your weekend goes well. I think the rain will hold off, which will be good for the startup on Monday (if everyone shows up on time, otherwise the rain'll have to hold off until Wednesday, I suppose). By the way, I should be freed up in about 3 weeks, and I'd like to have it off.

Questions

1. After reading this email, your boss will _____.

   a. think you want to be congratulated for staying on schedule

   b. think you want to be thanked for working overtime

   c. understand that you have things under control

   d. not know for certain what you actually want

2. To understand your email, your boss _____.

   a. should reread the email several times to find out what you want

   b. should highlight the main points of your email

   c. will probably have to guess what you want

   d. will probably have to call you or send an email asking what you want

3. If you were the boss and received this email, you should probably _____.

   a. tell the employee never to send you an email again

   b. suggest the employee work on his or her writing skills, perhaps by taking a course

   c. tell the employee that, from now on, to give you information only in person

   d. send an email in reply, telling the employee to stop wasting your time and to send you another email that's more to the point

4. You wrote that "a few things" were worrying you. After reading your email your boss will know what all of those things are.

   a. True

   b. False

5. After reading this email, your boss will know what day you want to start your leave and how many days off you want.

   a. True

   b. False

Activity 6: Paying Attention to Details

Details count, especially in construction work. Work with three or four of your classmates. Read each of the following situations and figure out what, if any, details are missing.

Team Members

1. _____   2. _____

3. _____   4. _____

Questions

1. For a fire to start, three things must be present in the same place at the same time: heat and oxygen.

   Are all details present?   ❑ Yes   ❑ No

   What's missing? _____

2. You must follow four steps when using an orbital sander: Get a good grip on the tool before you turn it on. Let the sander come up to full speed before you set it against the surface to be sanded. When finished, don't set the sander down right away.

   Are all details present?   ❑ Yes   ❑ No

   What's missing? _____

3. You must never wear contact lenses while welding. The ultraviolet rays might dry the moisture beneath the contact lens, causing it to fuse to your eye. If you then try to pull out the contact lens, you could rupture your cornea and go blind.

   Are all details present?   ❑ Yes   ❑ No

   What's missing? _____

4. You will see four types of fire extinguisher labels: Class A (green), Class B (red), Class C, and Class D. Each stands for a different type of combustible fuel: A for ordinary combustibles, B for liquids, and C for electrical equipment.

   Are all details present?   ❑ Yes   ❑ No

   What's missing? _____

5. To get to the new jobsite, travel on Interstate 95 for 10 miles. Take the exit marked Glades. Turn at the end of the ramp, and you'll see the new site in a little while.

   Are all details present?   ❑ Yes   ❑ No

   What's missing? _____

Appendix A: Finding Information Fast in Books and Manuals

Suppose you want to find safety information on fire extinguishers in a 400-page manual. Flipping through all those pages would take a long time. (If there really were a fire, that would not be the time to be looking in a manual.) To quickly find the information you need, remember to use the cues discussed in the *Reading on the job* section of this module. To look up a topic quickly, you would use one or more of the following cues: table of contents, section headings, index, and glossary.

You will find the table of contents at the front of the manual. In this case, as you look through the table of contents, you'd be looking for a chapter about fires. An example would be:

You would then turn to page 35, which is the beginning of Chapter 12. Page through the chapter, looking for section headings about fire extinguishers, and look for a subhead that might read *Fire Extinguisher Safety*. But there's an even faster way to find the information. Go back to the table of contents and find the page number where the index starts. The index is the section that lists all the topics in the document, in alphabetical order. Going to the index, you'd look for the topics beginning with the letter F. You'd look for the specific topic *fire extinguishers*, and you'd find an entry like this:

That way, you could go straight to page 142 to find out about fire extinguisher safety and to page 145 to read about the types of fire extinguishers.

While you're reading the safety information on page 142, say you come across the word **accelerant** and it's in boldface type (like it is here). The use of boldface is another cue: in this case, it's likely an indication that the word is defined in a glossary, which is a kind of mini-dictionary many documents include. You can find the glossary by checking the table of contents; the glossary is usually at the very end (or sometimes, at the beginning) of a document.

The table of contents, section headings, index, and glossary are the most common cues to finding information quickly in a book or manual. Here are three other helpful cues:

- **Tables.** Tables commonly include such information as statistics, weights and measures, pipe sizes, abbreviations, machine tolerances, machine parts, and load specifications.

- **Figures.** Figures can be photographs, maps, and line drawings. Many books and manuals have a special table of contents that lists only the tables and figures in the document; some books have separate tables of contents for both tables and figures.

- **Section tabs.** Publishers sometimes add tabs to section dividers. The tabs project a little beyond the pages, with the section name printed on the tab sticking out from the document, allowing you to quickly spot and page to the section you need.

Appendix B: A Quick Guide to Reading Plans and Blueprints

Most blueprints aren't actually blue, but because these plans were once shown as white lines on a blue background, they're still called blueprints. Learning how to read blueprints takes time, and you will learn more about this topic as you advance in your construction career. For now, here's a short guide to how to use them.

- **Know what type of blueprint you are looking at.** All blueprints include a title that describes the drawing. Some examples of titles are Roof Framing Plan, Floor Plan, Front Elevation, Finish Detail, Plumbing Plan, and Electrical Plan.

- **Figure out where you are in relation to the drawing.** Sometimes a drawing shows what a structure would look like if you were standing right in front of it (an elevation). To understand some drawings, like floor plans, you have to imagine that you have peeled the roof back from a structure so that you are looking down into the rooms. Some plans show you a side view of a structure. Some drawings show only a part (section) of a structure, and some drawings show a close-up view (detail).

- **Know where to look for the meaning of symbols and abbreviations.** Blueprints contain a lot of information, much of it communicated through the use of tiny symbols or abbreviations. A blueprint usually has a legend, which is a list of those symbols and what they mean. Abbreviations are spelled out on an alphabetical list. Legends and lists may appear right on the drawing or be included on separate pages in a set of plans.

- **Know where to look for the scale of the drawing.** You will usually find this information next to or below the title of the drawing.

- **Know where to find additional information about the plans.** All sets of plans include general notes with additional information on how the structure will be built, what materials will be used, and what building codes will be followed.

- **Know what to do if you have a question about a blueprint.** If you can't figure out a drawing or if something looks wrong to you, always mention it to your supervisor. If you find something that turns out to be an error in the drawing, your company will have a process in place to deal with it.

TOOLS FOR SUCCESS: CRITICAL SKILLS FOR THE CONSTRUCTION INDUSTRY

The speed, flexibility, and convenience of email make it a popular tool in business for communicating. The sender can attach text and image files and send them in far less time than it would take to type, photocopy, assemble, and distribute a paper memo. *Figure 7-3* is an example of a typical email.

Figure 7-3. Sample email

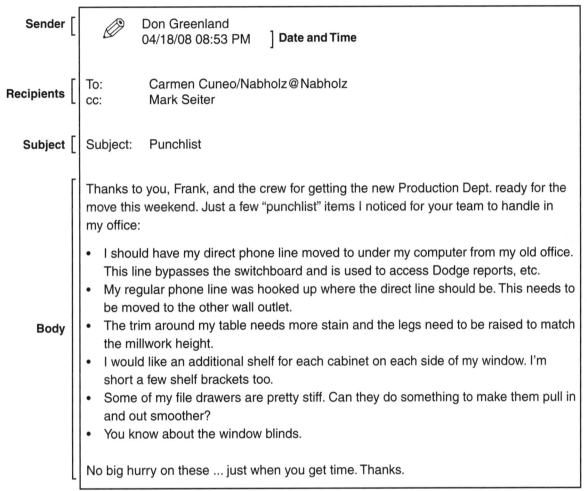

Source: Nabholz Construction Corporation, Conway, AR

The email above has five parts:

- The sender—in this case, Don Greenland is the name of the person who sent (and presumably wrote) the email.

- The date and time—in this case 4/18/08 08:53 p.m., which means the email was sent by Mr. Greenland on April 18, 2008, at 8:53 in the evening.

- The recipients are the people the email was sent to—in this case, Carmen Cuneo and Mark Seiter. When you list a name in the box next to cc (carbon copy), it often means the person is not the primary recipient of the email but is being included on the distribution list as a courtesy.

- The subject—in this case, Punchlist, which is the writer's way of summarizing what the email is about in one word. Subjects are usually only a word or short phrase.

- The body, or the content of the email. In a business email, this usually includes a quick greeting, a couple of brief points (often in bullet form, as in this example), possibly some type of request, and then a cordial conclusion.

There are many advantages to using email, but there are drawbacks as well. Here are some tips on how to use emails wisely:

- **Be professional when you compose your email.** This doesn't mean you have to worry about perfect grammar or sentence structure, but you should make sure that your email is like anything else you'd write on the job: clear, concise, and to the point. You also don't want to use such things as emoticons (such as smiley faces); you don't want to type in all capital letters (some people consider that to be the email version of shouting and therefore rude); and you don't want to use a lot of abbreviations (like IMHO, LOL, or AFAIK) your readers might not understand. (Do you know what these mean?)

- **Email is not private.** Anyone, including your boss, can read an email. In addition, people who receive an email from you can forward that email to practically anyone, as long as they know that third-person's email address. So never write or attach anything that could possibly be seen as insulting, overly personal, rude, or pornographic. Unless you know that you have a secure connection, avoid putting any personal information in an email. Many people have gotten into serious trouble for doing these things.

- **"Send" cannot be undone.** Once you press the Send button or command, your email is gone and can no longer be edited, so carefully review what you write before you hit your keyboard key or click your mouse. Make sure dates and amounts are correct. Make sure your description on the subject line is reasonably clear and not confusing. Make sure you're sending it to the right people. Don't rely entirely on your email spell checker; if you type *furnace flu or flew when you meant flue*, the spell checker won't help you.

- **Don't hide behind email.** email is not a substitute for face-to-face interaction. Never send bad news through email, and don't use email to avoid your responsibilities. If you've made a mistake, don't email your boss about it; show respect by communicating this in person.

Managing Stress on the Job

"Don't let your mind bully your body into believing it must carry the burden of its worries."

– Astrid Alauda,
psychologist and author

"I try to avoid stress—it makes me feel like I'm rubber-stamping all my organs 'Urgent'."

– Berri Clove,
writer

Introduction

We all have demands on our time. Life is a constant tug-of-war between the things we like to do and the things we need to do. Balancing our wants and our responsibilities is a challenge; it makes life interesting, but it makes life stressful, too.

Stress can be caused by actions we take or by things that happen to us; it can come from things we expect or things we don't expect. Major events are significant sources of anxiety, but even an apparently insignificant event—for instance, you wait in a checkout line for 15 minutes and then the clerk closes down the checkout station just as you are about to go through—can raise our stress level. It is not just hardships like money worries, or reversals like getting laid off from your job that cause stress; preparing for or participating in events that bring us great happiness, such as getting married or becoming a new parent, can raise our anxiety levels.

There is nothing wrong with feeling under a bit of pressure occasionally; it is unavoidable and can be beneficial. Too much stress, however, can hurt your concentration or make you sick. You may not be aware of how much stress you are under on a daily basis, yet its effects could be accumulating. The good news is you can identify stress and its causes. In this module, you will learn some techniques to help manage stress and stay healthy.

> "I try to take one day at a time, but sometimes several days attack me at once."
>
> – Jennifer Yane, author

Recognizing stress

Stress doesn't always announce itself. You may think you are moving through your daily, weekly, or monthly routine, going to work, going out with friends, with everything under control. Stress could be building inside you and seeping out in ways you do not see but that your closer acquaintances do. A friend might observe that you are acting differently. A co-worker might wonder why you seem to be more distracted at work these days. Your boss might notice you are less attentive than you used to be.

Any of the following symptoms could be a sign you are under more stress than you should be; the more of these you exhibit, the more likely it is you are under too much stress:

- You can't sleep, or you sleep too much.
- You have trouble concentrating or remembering things.
- You are tired all the time.
- You often feel depressed, hopeless, or anxious.
- You frequently have nagging headaches or chest, back, or stomach pain. (These can also be signs of an underlying health problem; if these persist, you should go to the doctor.)
- You are irritable around people at work or your friends and family.

Tips for managing stress

People usually feel the most stressed-out when they believe they are victims of circumstances. Taking positive action, therefore, is one effective way of coping with stress. If you have been feeling under a lot of pressure recently, try one or more of the following techniques.

Address those issues that cause you stress. If you know what is causing you to be stressed, you can reduce or eliminate stress by working on its sources. Certain causes of stress, whether on the job or elsewhere, can be dealt with pretty easily. For instance:

- Are you frustrated because the soda machine at work always seems to be out of change or your favorite drink? Try bringing in your own sodas in a cooler.

- Do you hate waiting in line? Instead of muttering about how slowly the line is moving, pass the time by reading a book or magazine while you are waiting, or listen to your MP3 player. If you always seem to be in line at the depot waiting to pick up job supplies, call ahead with your order.

- Do you think there's too much to do at work and not enough time to do it? Come up with a plan to help even out your workload, and show the plan to your supervisor so that you'll still make your deadlines.

- Are you dreading an upcoming job interview? Prepare as much as you can, but also try to get your mind off it by exercising or doing something you enjoy. Then the night before, get a good sleep. Focus on how well you are going to do, and not on how poorly you are afraid of doing.

Keep tools and equipment in good working order. Few things cause more instant stress than when a tool or machine breaks at a critical time. You can't prevent every breakdown, but you can avoid most of them. Periodically check your tools and machines. Perform the manufacturer's recommended maintenance. When tools or equipment do break down, repair them or find a replacement right away, even if it is not a pressing matter, so that you will have them ready for use when you really need them.

Be as neat as the job allows. A messy work area makes it difficult to find what you need, and that can cause you frustration and stress. It can also create stress for the people around you if they have to find something in the middle of all your clutter. A messy work area is also unsafe. To reduce stress and stay safe, put your tools away after you've used them or at the end of the day; keep floor and work surfaces clear and free of debris such as wood and metal shavings; and dispose of waste materials promptly and properly.

Plan ahead. Before you start a task, gather all the necessary tools and supplies and put them in one place. You'll have everything you need at hand once you start, and you won't be frustrated by having to stop work to get some tool you discovered you need or to obtain more supplies. If you have to pick up supplies, call or fax your order ahead of time so it will be ready when you arrive to pick it up.

Do the hardest tasks first. Whenever possible, do the hardest part of the task first. The longer you wait to do something difficult and the more time you spend brooding about it, the more stress will build. You may not be able to do this with all your projects, but whenever you can, get the tough stuff done first.

Eat right and take care of yourself. Many people smoke, eat junk food, take drugs, or drink alcohol to settle down. These habits may have a temporary calming effect, but that does not last long. Meanwhile, the cumulative effect of unhealthy practices will make your health worse down the line, and your overall stress level will be higher than ever.

Exercise. Many construction workers think they get enough exercise on the job. Obviously, construction work can be strenuous and physically demanding, but that doesn't mean construction work alone will keep you in shape. Most construction tasks do not give you much aerobic exercise, which includes running, swimming, or riding a bike. (Think Lance Armstrong instead of Arnold Schwarzenegger.) Aerobic exercise is a great stress reducer; it gets your heart rate up, helps to release tension, and helps you to sleep better. It will also keep you in shape for the job.

> "Just can't live that negative way; make way for that positive day."
>
> – Bob Marley, musician

Get rid of negative thoughts. When people get down on themselves, they are not doing themselves any favors. Worry or low self-esteem can sap your strength and desire, and it can lead to a great deal of anxiety. That kind of stress can be readily controlled by focusing on the things you are good at instead of worrying about the things you don't think you can do. Be proud of what you can do and instead of brooding over what you cannot do, think about how you can improve.

Be flexible. Don't expect everything to go perfectly. If you do, you are bound to be frustrated; things rarely go 90 percent according to plan, much less 100 percent. If you can roll with the punches and adapt, you'll experience much less frustration and stress.

Manage your money wisely. Many people point to the lack of money as their number-one worry. Money is a necessity, but worrying about it doesn't pay, literally or figuratively. If you manage your money well, you'll manage the worry:

> "Change is inevitable, except from a vending machine."
>
> – Anonymous

- Make a budget so you'll know how much money you need to pay your bills each month.

- Before you buy anything, ask yourself if you really need it or just want it.

- Limit your credit card use and try to pay off or pay back your balances as soon as you can; the interest and penalties that credit card companies charge for not paying your bill on time or for carrying a large balance are paycheck killers.

- Shop for bargains and sales.

> "It's not money that's the root of all evil; it's the lack of it."
>
> – Anonymous

- Don't go out to dinner or happy hour as much; pack your lunch instead of buying it.

- Try to save some money from each paycheck, even if it is only a couple of dollars. It is the habit, not necessarily the amount, that is important.

- Remember, if you are a good employee you'll soon be earning more, which means you can spend more.

Keep work stress separate from home stress. Sometimes it is hard to separate the two. However, try to leave work problems at work and home problems at home. Remember that work is only part of your life.

Hang on to your sense of humor. Laughter is an instant muscle relaxant. You may not be able to laugh your troubles away, but laughter can at least keep them at a distance for a while, where they won't look so large. A healthy sense of humor can deflate a lot of stress.

> "Laughter is an instant vacation."
>
> – Milton Berle, comedian

Breathe deeply. When you are stressed, you tend to breathe in short, shallow breaths, which increases muscle tension. Try taking several slow, deep breaths instead. Imagine you are flooding your body with good healthy air and driving out all the poisonous tension. It is hard to feel edgy when you are breathing deeply.

Improve your appearance. It is true that when you look good, you feel good. So when you are feeling down, spruce yourself up.

Do something for someone else. A good way to take your mind off your own problems is to help someone else with theirs. Volunteer to help others in your community. Do something nice for a friend or family member. You'll get a lot of satisfaction from your efforts, and you'll see your own problems in a new light.

Health Care

Many of the warning signs of stress are similar to the warning signs of other health problems. You should never ignore potentially serious symptoms because you don't feel you can pay to get them checked. Doctors are expensive, though; what if you don't have health insurance and don't have the money to see a doctor? There are steps you can take to limit the cost of health care if you need it:

- Check with your company's human resources department to see if you are eligible for some kind of health care. They may recommend resources you can use, or suggest free or inexpensive health-care clinics in your area.

- Check to see if the county you live in has a public health department that offers free or discounted medical services.

- Many uninsured people may qualify for state-run insurances like Medicaid; check to see whether you are eligible.

- Unless you have a serious or life-threatening condition, don't go to the emergency room. It's very expensive, and there are other places you can go if you've got a bad sore throat or a sprained ankle. An urgent care facility, for instance, is much cheaper than the emergency room.

- If a doctor decides you need a prescription, ask if the doctor has any samples. If the doctor writes you a prescription, shop around the pharmacies in your area: a prescription that costs $60.00 at one drugstore might only cost $15.00 at another. Ask the pharmacist if the $15.00 prescription is available as a generic. Generic medicines are the same as brand-name medicines, except generally much cheaper.

- Take care of yourself. That means eating healthy food and maintaining a proper weight, exercising, not drinking too much alcohol (and *never* drinking and driving), avoiding drugs, getting plenty of rest, and wearing your seatbelt.

- Don't start smoking, and quit smoking if you do smoke. Quitting smoking can be hard, but there are a number of ways you can get help to stop. Remember, not only does smoking ruin your health, it's an expensive habit. A pack of cigarettes is $4.00–$5.00. If you smoke a pack a day, that's $120.00 to $150.00 a month. Just think of all the excess cash you'd have if you quit smoking.

Stress busters

A good way to keep stress in perspective is to find time to do something you enjoy, such as one or more of the following activities:

- Participate in a sport or exercise.
- Do volunteer work.
- Take up a hobby.
- Relax your mind with a favorite book, movie, video, or CD.
- Relax your muscles by taking a nap or a hot bath.
- Play with your children or your pet.

Example Scenarios

Each of the following examples presents a stressful situation followed by suggestions for dealing with it. Keep in mind these are suggestions; because everyone reacts to stress differently, everyone handles stressful situations somewhat differently. Before you read the suggested stress busters for each situation, ask yourself how you would deal with that scenario.

> **Example 1:** You and a co-worker are building a wall on a hot day. Sweat keeps dripping into your eyes, and you are having trouble holding your tools because your hands are so sweaty. Mosquitoes keep biting you. You are scheduled to take a break, but you are a little bit behind, so you skip your break and keep working. Your co-worker accidentally steps on your foot and you start shouting and cursing.

How would I deal with this?

_____.

Suggested stress busters: Take that break. It will give you time to drink some water, wipe off the sweat, and apply insect repellent. Tie a bandanna on your forehead to keep the sweat from your head out of your eyes. Wipe your hands with a cloth so you can handle your tools safely. Apologize to your co-worker and maybe make a joke about how you are the mosquitoes' favorite meal. Then get on with the job.

> **Example 2:** Your co-worker constantly jokes about how you talk. At first, you just shrugged it off. Lately, though, the comments are coming thicker and faster, and you've been getting pretty irked. Today, your co-worker crossed the line, poking you in the ribs to see if you were listening to the jokes.

How would I deal with this?

_____

Suggested stress busters: Take a deep breath, count to 10 or 20, and if you still feel like you are about to blow, walk away. A little break will help you to clear your head of the anger you naturally feel. You are not backing down; you are acting maturely and like a professional, unlike your co-worker. Once you've settled down, talk to this person. Calmly say this kind of joking is unprofessional and disrespectful and gets in the way of the job—everyone's job, not just yours—and that it is time to stop. If you don't get satisfaction or if the co-worker cops an attitude, drop the subject for now and think about the next steps you can take to resolve the issue effectively yet professionally.

Example 3: You've got your share of money woes. You owe three creditors and have only enough money to make the minimum payment to two of them. Meanwhile, every Friday your co-workers go out for drinks and you spend $25 to $30 each time you join them. You don't drink as much as most of the others do, but since the group always divides the bill up evenly, you feel a little gypped. Still, you like going out because it is a good way to unwind, which you need. You are feeling a bit guilty that you aren't spending Friday evenings with your child, and you'd like to buy the kid a nice toy to make it up. But there's rent to pay, gas to buy, and the dishwasher needs repair. Then there's this motorcycle for sale in your neighborhood, the kind of bike you've always dreamed of owning. Because of all this, you are considering taking on a part-time job, even though your current construction job is very tiring.

How would I deal with this?

Suggested stress busters: Talk to someone you know and trust and who seems to handle money well, and ask for some advice. Or, consult with a credit counselor who can help you figure out a more workable payment plan (that is, if the service itself doesn't cost too much). In the meantime, stop going out for Friday happy hour. Spend that time with your child instead; you'll probably have more fun, it will not set you back 30 bucks, and it is a great way to unwind. Your child would much rather have your company than any toy, and that should make you feel good. Use just one credit card, as infrequently as you can, and pay the others off and then get rid of them. Put aside for now the idea of buying the motorcycle; there will always be others for sale once you've got more money. If there is anyone at work who knows about appliance repair, ask that co-worker how you could get your dishwasher fixed inexpensively.

Don't Fret Over Small Things

Don't sweat the small stuff, the saying goes. Yet people seem to get upset a lot by some pretty insignificant incidents. Waiting in line for 15 minutes? Just think of the folks behind you who'll have to wait 16 or 17. Some creep cut you off in traffic? Settle down, it wasn't a personal insult by that driver. It just so happened it was your car that got cut off; it could just as easily have been someone else's. Bought a new drill and the box didn't have all the bits it should? Take the item back and get another, it may take an hour or so. If you're 22 years old, you've been alive for about 193,000 hours; if you're 32, you've been alive more than 280,000 hours. What's another hour? When you realize that so many of the things you stress out over aren't worth it, your overall stress level will go down, possibly dramatically.

Making the most of your time

People with stressful lives often think, "If only I had more than 24 hours in a day, I could do everything that needs to get done." Well, they don't, and if they did, they'd probably just add more to their schedules and they wouldn't feel any less stressed. To get everything done, you don't need more time; you just need to manage that time better. You will experience less anxiety if you make a time-management plan, if you don't put things off and don't waste time, and if you look for opportunities to save time as you go through your day.

- **Make a plan and stick to it.** Meeting a construction project's overall deadline depends on all workers meeting their individual deadlines. To make sure you keep to your schedule, look around the jobsite before you leave for the day and think about the most important things you will need to do on your next shift. Then come up with a quick plan on how to get them done safely, well, and with a minimum of stress. Once you get home, write down the plan, if you wish. You will find it helpful to have a written plan; when you get home from work, you can cross off what you got done during your shift and see where you stand in terms of your overall schedule. As you come up with your plan, ask yourself, "What are the most complex tasks and what are the most pressing deadlines I face?" Using your plan, you can manage your work schedule so that you will leave enough time for the most complex and pressing tasks, by knowing what your priorities are each day when you arrive at work. (Remember, though, your boss's orders come first; if they conflict with your plan, do what your boss says, and pick up with your plan later.)

- **Don't procrastinate.** When you procrastinate, you put off doing something you don't want to do by saying, "I'll get to it laeter." The problem with procrastination is that if you do not want to do something now, you will probably want to do it even less later—and you will have less time to do it then. That's a recipe for instant stress. You are not going to be thrilled by every task you are assigned, but those tasks have to be done regardless, and putting them off will just make it harder on you. It is helpful to do the things you do not like to do first thing in the morning, if your overall schedule allows it, which leaves the rest of the day for more pleasant work.

- **Don't waste time.** It is easy to get distracted on the job by talking with your co-workers about your weekend plans, about your problems at home, about how excited you are about your relatives coming to visit, about how the Lakers or the Bears are doing. A little chitchat is fine (no one expects you to be some kind of robot), but it can eat up an amazing amount of time if you let it. Limit the small talk; otherwise, you might end up stressing out at the end of your shift trying to complete something you should have done hours ago, except that you were too busy shooting the breeze.

- **Look for ways to use time wisely.** Say your workday includes the following tasks: going downtown to pick up some permits, hanging some doors, picking up some supplies from another worksite, taking some bags of garbage to a nearby dump, and shingling a roof. In what order would you perform these tasks? As we have seen, it is usually best to do the most difficult or time-consuming tasks first, but that may not always be possible. When planning your day and deciding in what order to do these jobs, you could ask yourself the following questions:

  - How big is the roofing job? Is it a hot day? It is easier to put on a roof when it is cooler. Are other people relying on you to help with the roofing at a certain time?

  - Where are the garbage dump, the other worksite, and downtown in relation to one another? Could you hit them all on the same trip?

  - Is there enough room in the truck for both the supplies and the garbage, so that you could pick up the supplies before you drop off the garbage? Or do you have to drop off the garbage first to make room for the supplies?

- If you pick up the supplies before the permits, and then you park and leave your truck unattended to get the permits, would the supplies be safe from potential theft?

- Are there people who can't start their jobs unless you hang the doors first? Unless you get the shingling done first? Are there people who can't do their jobs until you pick up the supplies?

Let's say it is a hot day, the roofing job is large enough to need a crew and the crew is counting on you to be available during the morning, there's not enough room in the truck for both the supplies and the garbage, the supplies will be safe when you go in to get the permits, downtown is closer than either the garbage dump or the supply depot, and the doors don't have to be hung until the end of the day. Here's an ideal way you could plan the day so that you get everything done with a minimum of stress:

- Shingle the roof in the morning when it is coolest and when you are needed.

- Load the garbage bags into your truck. Drop off the garbage on your way to pick up the supplies.

- On the way back with the supplies, stop off downtown to pick up the permits.

- Once back at the site, unload the supplies and the permits, and then hang the doors.

Summary

Stress may be hard to avoid at times, but it can be managed. The first step is to know the warning signs of stress; they include sleeping too little or too much, irritability, lack of energy, trouble concentrating, and sadness. To reduce stress, take control of the situation by addressing the issues that are causing you anxiety. You can also manage stress by good planning and proper work habits, taking care of yourself, managing your money wisely, focusing on your strengths rather than your perceived weaknesses, keeping a sense of humor, and relaxing overall. If you are stressed because you never seem to have enough time, come up with a time-management plan, don't waste time, and don't put off until tomorrow what you can do today.

Here's a quick quiz that allows you to apply what you've learned in this module. Select the best possible answer.

1. One of your co-workers says to you, "Wow, you look really stressed out." What might you be doing to make your co-worker think this?

   a. Acting distracted and not focusing on the job.

   b. Drinking a lot of coffee and chain-smoking.

   c. Reading the newspaper instead of working.

   d. Looking off into the distance while biting your nails.

2. A row of coat hooks is located near the break area at work. There aren't enough hooks for everyone, so workers just double or sometimes triple up when hanging their jackets on the hooks. At day's end, workers waste time cursing and looking for their jackets, and some jackets always fall on the ground. It is a small problem, but it annoys people and causes some stress at the end of the day. To solve this problem in a positive way, you should _____.

   a. get to the area before anyone else and remind everyone to play nice while they get their jackets

   b. put up a sign that reads, "When all the hooks are filled, workers will have to find another place for their jackets"

   c. Stay late one night and remove all the hooks, because that'll be a great joke to play on everyone and will do wonders to lower the overall stress level

   d. install more hooks

3. You are in the middle of ripping clapboards for a custom siding job when your circular saw jams and stops operating. What is the least stressful way to deal with this problem?

   a. Criticize yourself for failing to maintain your saw, and feel rotten the rest of the day so that you'll remember to take better care of it in the future.

   b. Take your safety goggles off and throw them in anger as far as you can, then go take a break and practice some calming breathing techniques.

   c. Fix the saw or find a replacement saw, and remind yourself to do a better job of maintenance in the future.

   d. Tell your supervisor that the saw is broken and that you feel rotten, and that you don't think you can stay on schedule.

4. You've had a tough day on the job. You're exhausted and a little bit behind schedule, so you are feeling really stressed out. You don't want to take out your frustrations on your family. What is the best way to deal with your stress?

   a. Go to the gym, if you belong to one, and work out, or go jogging or take a long walk.

   b. Go for a three-hour drive after work and think about all the lousy things that have ever happened to you.

   c. Ask your boss if you can work overtime that evening because that would be better than going home, where you might get into an argument.

   d. Complain to your boss or a co-worker about how no one understands you at home.

5. You're worrying a lot about your job. In particular, you're scheduled to do some detailed woodwork next week, and it is going to take a lot of time and skill. You're tired, and you're afraid you're going to make a lot of mistakes and get fired. Then how will you pay your bills? What if they re-possess your car for nonpayment? Meanwhile, you're trying to stop smoking. What is the best way to deal with this situation?

   a. Go out for a nice dinner, have a few drinks and smokes at the bar afterwards, and buy yourself a carton of cigarettes to get you through the week, since they're cheaper that way.

   b. Tell yourself you'll do fine at your job because you're skilled and prepared, then spend your weekend resting, relaxing, and doing something you enjoy instead of worrying. Buy some gum you can chew during the next week to help you relax.

   c. Tell your boss you are not up to doing this job, that you're preoccupied with money worries and trying to quit smoking, and ask that the task be assigned to someone else.

   d. Eat a lot and drink a lot of coffee so you'll have lots of stored-up energy for next week, and start smoking again (it'll only be temporary, until the woodworking is done) because it helps you to relax.

6. You are scheduled to do roofing today. You arrive at work in a great mood, because you really enjoy roofing and it is a good day to do it: nice and cool. However, your boss tells you that you're needed to do some trenching work instead. You absolutely despise trench work; you'd rather sweep the floor. What is your best course of action?

   a. Recognize that the situation is temporary, be flexible, and do a good job.

   b. Tell your boss that you get uncomfortable and stressed out working in closely confined spaces like trenches and to please assign someone else to the task; offer to sweep the floor instead.

   c. Tell your boss that the orders will play havoc with your carefully prepared work schedule, but go do the trench work after making sure your boss knows how reluctant you are to do it.

   d. Do the trench work, but do a sloppy job, and your boss will get the point.

7. You earn a good salary, but unexpected expenses pile on more bills than you can pay. The best way to manage the stress caused by your money problems is to:

   a. Join a local gym and put it on your credit card, because daily exercise is the best way to reduce stress.

   b. Talk to a credit counselor to find a manageable way to pay your debts, while swearing off happy hour and big purchases for the near future.

   c. Go slam the vending machine a couple of times to take out your aggravations and, who knows, you might get a little pocket change out of it.

   d. Volunteer to help the needy, so that you get your mind off your money problems by being around people worse off than you are.

8. Your boss, who is normally easygoing and reasonable, is getting a divorce and has been critical of your work lately. You recognize that your boss is letting stress at home affect the job. Now you're feeling stressed, too. What is your best course of action?

   a. Ask your boss to re-read this module, and suggest that home stress should not be brought to the workplace.

   b. Tell your boss's supervisor that the quality of your work is really suffering because the boss is bringing personal problems to the job and taking it out on you.

   c. Recognize that the situation is temporary, don't take the boss's criticisms personally, and focus on doing a good job.

   d. Sympathize with your boss by saying you know how stressful getting a divorce is because your parents got one (even though they didn't).

9. A co-worker who has a foul mouth and a rotten temper curses at you for no reason. You feel angry and insulted and want your co-worker to know that you should be treated with respect. What is your best course of action?

    a. Yell back at the co-worker using even fouler language, which will help you let off steam and allow you to make your point without having to resort to more dramatic means.

    b. Count slowly to 10 before you say anything, and walk away if you're still angry.

    c. Let off steam by walking away, and then fantasize about slashing the tires of your co-worker's car.

    d. Make fun of your co-worker, because laughter is the best medicine.

10. Your boss tells you that you must accomplish the following four tasks by the end of the day:

    a. Sweep the floors of the house your team has almost finished building.

    b. Hang four doors in the house.

    c. Drive across town to pick up supplies the team needs by the end of the workday.

    d. Place a large order for lumber.

11. Assuming that none of the other workers' schedules will be affected by any of these tasks, in what order should you do them?

    a. Order the lumber, hang the doors, sweep the floors, pick up the supplies.

    b. Hang the doors, pick up the supplies, order the lumber, sweep the floors.

    c. Pick up the supplies, hang the doors, order the lumber, sweep the floors.

    d. Sweep the floors, hang the doors, pick up the supplies, order the lumber.

Individual Activities

Activity 1: Eliminating Your Time Wasters

We've seen how important time management can be. One way to manage your time better is to identify what causes you to waste time. These can be personal habits such as being disorganized or trying to get too much done at once, or activities or tendencies such as staying up too late or talking too much at work. Some time wasters lead to others; for instance, if you are disorganized, chances are you spend a lot of time looking for things. If you stay up late partying on a work night, you may be tired and hung over the next day, which means you work more slowly and fall behind schedule.

The following list includes some common time wasters. Check any that apply to you, and then develop an action plan for dealing with them. An example of an action plan to deal with procrastination is provided.

Check Here	Time Waster and Result
❏	Being disorganized, which means you waste time looking for misplaced items
❏	Trying to do too much in too little time or all at once, which means you don't do anything as well as you can and as well as you should
❏	Getting up too late for work, which means you rush off unprepared for the day
❏	Being too much of a perfectionist, which means you have less time for other tasks
❏	Spending too much time chatting with co-workers, which means you have to finish up your responsibilities all at once at the end of the day and therefore don't do any of them properly
❏	Staying up late partying, which means the next day you are tired, hung over, or both, and you can't work as quickly and as efficiently as you need to
❏	Failing to plan ahead, which leads to being disorganized
❏	Allowing too many interruptions to your workday, which means you have less time to get things done
❏	Failing to listen to or read instructions, which means you have to do the task all over again when you should have moved on to something else
❏	Spending too much time fiddling with small, unimportant tasks, which means you have less time to deal with the important stuff
❏	Working too slowly, either because you are trying to attain an unrealistic level of perfection or because you are tired from lack of rest from too many long nights, which means you fall behind schedule and are putting the overall project at risk
❏	Procrastinating, which means you'll have less time to do those things you don't like when you can no longer put them off

Sample Problem: Procrastination Sample Action Plan

1. Divide a big job into small tasks so I don't feel overwhelmed.

2. If possible, do the hardest tasks or tasks I don't like first to get them out of the way.

3. Give myself deadlines for getting each task done.

4. Reward myself for completing the tasks or withhold the reward if I don't complete them.

My Time Wasters Problem: _____

Action Plan: _____

_____

Problem: _____

Action Plan: _____

_____

Problem: _____

Action Plan: _____

_____

Activity 2: Practicing Relaxation Techniques

When people feel stress, they often take shallow breaths. Shallow breathing, however, tends to make stress worse. The trick is to breathe deeply instead, which helps your muscles to relax, reducing your overall stress level. In this activity, you will practice doing these exercises. You might think these exercises are silly and you might be embarrassed to try them, but these exercises are good ways to reduce stress; they really work. (Once people see how relaxed you are, they might start asking you to show them how to do the exercises!) Try one or more of the exercises and then answer these questions:

1. Did you feel more relaxed after trying the exercise?

2. Can you think of a recent situation where these exercises could have helped you relax?

Deep-Breathing Exercise

This exercise will help you get more air into your lungs, which will help relieve stress in your shoulders and back. At first, you may feel more stress than relaxation, but once you learn the exercise, it will work for you. Repeat this series of steps five times for maximum effect.

1. Sit up straight in a comfortable chair.

2. Place one hand on your stomach and the other hand on your chest.

3. Breathe in slowly through your nose. When you breathe in, you should feel the hand on your stomach rise before the hand on your chest rises. Practice until this happens. As you continue to inhale, you will feel your shoulders raise a little as your lungs fill with air.

4. Hold your breath for five seconds.

5. Slowly exhale. As you exhale, your shoulders will drop.

Muscle-Relaxation Exercises

These exercises will help you relax when you feel tension in your head, jaw, arms, shoulders, neck, or legs. In each exercise, you will first increase and then relieve the tension. Repeat each exercise five or ten times, or until you feel the tension disappear. Focus on doing each exercise smoothly. Don't forget to breathe!

Forehead: Close your eyes, and raise your eyebrows as high as you can. Hold for five seconds and relax.

Jaw: Bite down and pull down the corners of your mouth in an exaggerated frown. Hold for five seconds and relax.

Arms: While sitting up straight in a comfortable chair, push your elbows into the back of your chair just enough to feel tension in your upper arms. Hold for five seconds and relax.

Shoulders and neck: Pull your chin down into your chest as far as it will go. Hold for five seconds and relax. Take a deep breath to release the tension from your shoulders.

Legs: While sitting up straight in a comfortable chair, raise your legs and point your toes. Hold for five seconds and then lower your legs.

Activity 3: Creating Your Own Budget

As you've learned in this module, lack of money is a big source of stress for many people. You may be faced with an unexpected expense: medical bills, for instance, or a new refrigerator to replace the one that broke and can't be fixed. But you also need money for things you want: to purchase a new car or go on a nice trip, to take classes in a construction trade at the local community college, or to upgrade your set of tools. Maybe you want to save money to purchase your own home or condominium.

A great way to keep money troubles from stressing you out is to learn how to make and work within a budget. If you have never done this before, this activity will give you the tools and information you need. First, consider these tips:

- To figure out how much money you need to save each month for a future event, divide the total amount needed by the number of months in the time period you have selected. For example, you've determined you'll need $20,000 in 10 years. You need to save $167 per month ($20,000 ÷ 120 months = $166.66). If you invest your savings wisely, you can make money on your savings, and you'll reach your saving goal in less time.

- Keep up with credit card payments. Pay off as much of your balance every month as possible. Pay off your credit card balances entirely if you can. Don't charge too much on your credit cards. As noted, if you carry a large balance on your credit card from month to month, you can end up paying a lot in interest.

To complete your budget, follow these steps:

Step 1 Add up the numbers to get the totals for Categories 1, 2, and 3.

Step 2 Add the totals of Categories 2 and 3 together.

Step 3 Subtract that answer from the Category 1 total. At this point, it would be nice if you had some money left over for Category 4. You may or may not. Keep in mind that when you first start out, your budget will be a little tight. If you establish good saving habits, you will soon have a pool of money that will allow you to buy not only the things you need but also some of the things you'd like.

My Monthly Budget

Category 1: Money Coming in Each Month

My salary — $ _____

Any other income (from another job, for instance) — _____

Total Category 1 — $ _____

Category 2: Money I Must Spend Each Month

Rent or mortgage — $ _____

Insurance (life, health, home, and car) — _____

Utilities (phone, water, electricity, cable, gas, etc.) — _____

Car payments — _____

Car upkeep (gas, oil, maintenance) — _____

Food and toiletries — _____

Clothes and shoes — _____

Medical (co-payments, prescriptions, etc.) — _____

Other — _____

Total Category 2 — $ _____

Category 3: Things I Must Save Money for Each Month

Emergencies — $ _____

Retirement — _____

Education for myself or my children — _____

A home purchase — _____

Holiday purchases — _____

Other — _____

Total Category 3 — $ _____

Category 4: Things I Want to Spend Money On

Fun (eating out, entertainment, books, CDs or DVDs, hobbies, vacations) — $ _____

Other — _____

Total Category 4 — $ _____

Note

This budget shows common income, expense, and savings categories. You may have different categories on your personal budget. Just cross out anything that does not apply, and write in those things that reflect your life.

Saving and Investing

There are many different ways to invest the money you save; most of these involve some risk. A book on how to invest your money could be twice as long as this workbook. If you have some savings and want to invest them, talk to a reputable professional money manager before you do so. The bank where you cash or deposit your check may have a financial advisor. A lot of people have opinions on how to manage money, some of them quite foolish, so be careful who you talk to about this subject. If it sounds too good to be true, it probably is. Don't allow yourself to be scammed by get-rich schemes.

Group Activities

Activity 4: Turning Negative Thinking into Positive Thoughts and Actions

Negative thinking does more harm than good. It makes you feel less sure of yourself, which can lead to stress. Taken too far, negative thinking can lead you to believe all kinds of unfortunate and untrue things about yourself: for instance, that no one likes you or that you can't do anything properly. In this activity, you and three or four classmates will replace negative thinking with positive thinking and negative statements with affirmative ones. As you work through this activity, keep these ideas in mind:

- Focus on the behavior, not on the person.

- Mistakes may be stupid, but the person who makes one isn't. It is not a stupid mistake if you learn from it.

- You aren't born with the weight of the world on your shoulders. Ask for help when you need it; we all need help now and then.

- Be wary of thoughts that include the words *never* or *always*. You *never* seem to learn how to do anything? Does that mean you don't know how to drive, or walk, or talk, or tie your shoes? None of us were born knowing how to do any of that. You *always* get the short end of the stick at work? Does that mean you are the only one who has been reprimanded or who has had to work overtime?

- You may not be able to get rid of negative thoughts entirely, but you can shed most of them. One way is to take positive action; we usually feel better when we're doing something and not just letting things happen to us. If there's something you don't like about yourself, take action to improve it or ask yourself why you don't like that aspect of yourself in the first place. You may not like having to wear glasses, for instance, but lots of people wear them. Save up to purchase contact lenses, then, or just live with the fact you need eyeglasses. That's certainly no reason to hate the rest of who you are.

As an example, the first negative thought listed in the table has a positive response and affirmative action.

Team Members

1. _____   2. _____

3. _____   4. _____

Negative Thought	Positive Thoughts and Affirmative Actions
I can't believe that I did that wrong again! I'm so stupid!	*Example:* I am not stupid. I might just need additional training. I'll ask my supervisor about training opportunities.
No matter how hard I try, I will never be able to figure out how to do this job.	
I just know the boss is going to make me work overtime this weekend. Why is it always me?	
I am tired of the boss telling me what to do. No one has the right to do that.	
I don't think I would like being a boss, and besides, I am not good enough to do that job.	
I have to baby-sit everybody on this job. Co-workers always bring their troubles to me. Do I have a sign on my forehead that says, "All problems solved here"?	
None of the other workers like me. And I know the boss hates me. No wonder I can't get ahead in this job.	
The boss must stay up late nights thinking of ways to make my life miserable. I am always given the worst jobs.	
This job is too hard. I am not strong enough to do it on my own. Who do these people expect me to be? The Incredible Hulk?	
I am drowning in debt. I'll probably get kicked out on the street and starve. What have I done to deserve this?	
My co-workers are such worthless jerks. I've had to fill in for every one of them at one time or another.	

Activity 5: Busting Stress at Work

Here's your chance to create an action plan that you can take to the job. Work with three or four of your classmates to create your own workplace stress-busting plan, by answering the following discussion questions.

Team Members

1. _____ 2. _____

3. _____ 4. _____

Discussion Questions

1. What activity or exercise could I share with my teammates to help all of us deal better with stress?

2. If my co-workers like to go out drinking regularly and I don't, yet I still would like to go out with them, can I suggest some other activity we'd all enjoy and that didn't cost as much as bar-hopping?

3. Is there some minor yet annoying problem at work I could help solve, which might help to lower everyone's stress levels a bit?

4. If I convinced a bunch of my co-workers to sign up with me, could I get some kind of package deal at a local gym that would give us a group discount on membership?

5. Could I convince my co-workers to participate in bring-lunch-to-work days to save us all a few bucks?

6. Could I get some of my co-workers interested in volunteering for a community project, or doing something else to help out others, which could help reduce our stress and make us feel good in the bargain?

Thinking Critically and Solving Problems

"It's not that I'm so smart, it's just that I stay with problems longer."

– Albert Einstein

Introduction

Construction is a challenging, ever-changing field. There are many things you need to know and many topics you'll need to keep up with: new technologies and materials, and new procedures and methods. You'll regularly be presented with information and then have to decide whether this information is relevant to your job and valuable to you. That's where your critical-thinking skills come in.

During your career, you'll deal with a number of different challenges. To be a valued member of the team and a success in your construction career, you must be ready, willing, and able to handle these challenges. This doesn't mean you face a future full of stress and frustration; on the contrary, you'll grow to enjoy meeting these challenges—problem solving is among the most rewarding and interesting aspects of any profession, including construction. You'll get a lot of personal satisfaction when you successfully tackle a difficult problem, and that may also enhance your value to an employer.

To solve problems effectively, you'll need to have good critical-thinking skills. In this module, you'll learn about both critical thinking and problem solving, and you will get a chance to practice both.

Critical thinking

When you are thinking critically, you are doing two things:

1. Paying close attention to information that you read, hear, or otherwise absorb.

2. Evaluating what you've read or heard using your own knowledge and experience, as well as validating that information through research or further inquiry.

The goal of critical thinking is to help you make the best possible decisions about information you've received. When you think critically, you look past what is obvious by digging deeper into an issue, which gives you a more comprehensive understanding. When you are thinking critically, you don't jump to conclusions; you carefully consider an issue and then come to a reasoned and informed decision. (Note that the use of the word *critical* in critical thinking does not mean *negative* or imply that you are criticizing.)

Consider this example. You are choosing between two products. One is a popular brand name; the other is not so well known. Your company receives glossy brochures in the mail about the brand-name product; even after reading through the brochure twice, however, you can't find much substantial information about it. The brochure about the lesser-known product, on the other hand, seems to be more informative even though it isn't nearly as fancy. Experience has taught you not to buy something based solely on advertisements, however, even if the information is substantial. So you ask a well-respected co-worker for an opinion about the lesser-known product; this individual has great things to say about it. You're about to go ahead and buy the lesser-known product when you find out the co-worker's uncle is a top executive at the company that manufacturers it. Meanwhile, another co-worker, who has a lot of experience using both products, advises you to buy the brand-name one; you ask the co-worker a number of questions about the brand-name product and receive satisfactory, comprehensive answers.

Here are some examples of how you applied critical thinking in this example:

- You didn't make a snap decision to purchase the better-known product.

- You paid close attention to the information about the brand-name product (reading the brochure twice).

- You looked past the obvious—the glossy brochure about the brand-name product.

- You used your experience to dig deeper into the issue and not rely solely on advertisements, while researching your decision.

- You sought to validate the information about the less-known product by asking the opinion of a co-worker.

- You realized that a personal recommendation might not be completely objective.

- You validated the recommendation of another co-worker by asking a number of questions.

Separating Fact from Opinion

To think critically, you must be able to tell the difference between a fact and an opinion. If a statement can be proved true or false, that statement is a fact. An opinion, on the other hand, cannot be proved true or false because it is based on someone's point of view or personal interpretation of facts. You can get a fact wrong: someone who says "the moon is made of green cheese" is incorrect. On the other hand, an opinion cannot be proved true or false; someone who says "the moon is ugly" may be in the minority, but since there is no way to determine whether such a statement is true or false, it is an opinion. Different people can have different opinions on the same fact. If you live in a town where the nearest big city is 100 miles away, your opinion might be that's a long distance, while your co-worker's opinion might be that it's a short distance. Both of your opinions are about the same fact: the city is 100 miles away.

Using critical-thinking tools

Critical-thinking tools are a series of questions you use to evaluate information you've received about a situation or topic. These questions can be roughly defined as:

- **Tool 1:** How trustworthy is the source of the information?

- **Tool 2:** How does this information compare to what I already know about the topic or my experience with the situation? Does it make sense? Is it realistic?

- **Tool 3:** How do I feel about this information, and why? Is my opinion about the information somehow influenced by how I thought about the situation in the first place? Am I assessing the information independently and fairly?

- **Tool 4:** What other sources can I consult to validate whether the information is legitimate or not?

Read the following example—an advertisement that says you can learn construction math in a very short time if you use a company's proven and tested methods—and then see how these critical-thinking tools can be applied to the claims made by the advertisement: that you can learn a lot of math in just two days.

Tool 1: Can I trust the source of this information? You get information from many different sources: books and newspapers, the computer, television and radio, word of mouth. How can you choose between the sources that are valid and those that are less so?

A helpful rule of thumb is that a source is more trustworthy:

- The longer the source's track record for telling the truth (honesty and integrity)

- The greater the source's proven experience in an area

- The deeper the source's level of knowledge or expertise in an area

With these rules in mind, take another look at the ad. What is the source of this claim? Can you trust this source? Why or why not? What can you do to check the legitimacy of this source? Remember: a person or company can still be a valid source of information even if it doesn't have a track record. Just because you have never heard of XYZ Learning, Inc. doesn't mean the company is not a good one. People who are just starting out or brand-new companies might well possess the expertise they claim.

Tool 2: How does this information measure up against my own knowledge and experience? Your own knowledge and experience are important to critical thinking. Consider once again the ad and the claims it makes. Does it make sense that anyone could learn all those topics in two days? Perhaps some people could—people with a knack for math, for instance—but could everyone, even people with little background in math? Some people are better at math than others are; does the ad reflect that? Do you think it is possible to learn how to apply geometry to construction problems "instantly and without effort"? Can you learn something that's worth knowing "instantly and without effort"? Why does the advertisement urge you to "call today"? Couldn't it wait until tomorrow, or next week?

Tool 3: How do I feel about this information, and why do I feel this way? Sometimes you may have personal feelings about a topic or situation, which can undermine your ability to critically evaluate information about it. For instance, if you really dislike math courses, you could be tempted to take this "easy, fun-filled" course, even though all your critical-thinking skills may be telling you this offer is probably too good to be true. In other words, your desire to learn a difficult subject without making any effort might overcome your critical-thinking skills. When you think that your emotions, either good or bad, are getting the better of your judgment, put them aside and go over the facts again by using your critical-thinking skills, before you make any decisions.

Tool 4: What other sources can I consult about this information? Even when you apply your critical-thinking skills and put your emotions to one side, you might still be undecided. Maybe XYZ Learning, Inc. really has discovered an easy and revolutionary way to learn math. You could conceivably reach this conclusion through critical thinking, but you still might have reservations. That's when it is helpful to validate the information by asking someone who is knowledgeable about the topic or by reading more about the issue. In this case, if you were still uncertain about these claims, you could talk to a co-worker who is good at construction math. Maybe you have a family member who knows a lot about geometry and algebra. It's a good idea to talk to people with expert knowledge before making a decision, especially if you're still in doubt.

Keep these critical-thinking tools in mind as you work on the problems presented later in this module. These tools will enhance your ability to analyze and solve problems both at work and in your personal life. Remember: do not accept information at face value. Dig a little deeper. Make decisions based on facts, not emotions. Consult experts or people you trust.

Problem solving

The ability to solve problems is an important skill in any workplace. It's especially important in construction, where the workday is often not predictable or routine. In this section, you will learn a five-step process for solving problems, which you can apply to both workplace and personal issues. Let's look at the steps and then see how they can be applied to a job-related problem.

Step 1: Define the problem. This isn't as easy as it sounds.

Step 2: Think about different ways to solve the problem. There is often more than one solution to a problem, so you must think through each possible solution and pick the best one. The best solution might be taking parts of two different solutions and combining them to create a new solution better than the other two.

Step 3: Pick the solution that seems best and figure out an action plan. It is best to receive input both from those most affected by the problem and from those who will be most affected by any potential solution.

Step 4: Test the solution to determine whether it actually works. Many solutions sound great in theory but in practice don't turn out to be such winners. On the other hand, you might discover from trying to apply the solution that it is acceptable with a little modification. If a solution does not work, don't get rid of it immediately. Think about how you could improve it, and then implement your new plan.

Step 5: Evaluate the process. Review the steps you took to discover and implement the solution. Could you have done anything better? If the solution turns out to be satisfactory, you can add the solution to your knowledge base.

Applying the problem-solving process

Next, you will see how to apply the problem-solving process to a workplace problem. Read the following situation and apply the five-step problem-solving process to come up with a solution to the issues posed by the situation.

Situation: You are part of a team of workers assigned to a new shopping mall project. The project will take about 18 months to complete. The only available parking is half a mile from the jobsite. The crew has to carry heavy toolboxes and safety equipment from their cars and trucks to the work area at the start of the day, and then carry them back at the end of their shifts.

Step 1: Define the problem. Workers are wasting time and energy hauling all their equipment to and from the worksite.

Step 2: Think about different ways to solve the problem. Several solutions have been proposed:

- Install lockers for tools and equipment closer to the worksite.

- Have workers drive up to the worksite to drop off their tools and equipment before parking.

- Bring in another construction trailer where workers can store their tools and equipment for the duration of the project.

- Provide a round-trip shuttle service to ferry workers and their tools.

Note: Each solution will have pros and cons, so it is important to receive input from the workers affected by the problem. For example, workers will probably object to any plan (like the drop-off plan) that leaves their tools vulnerable to theft.

Step 3: Pick the solution that seems best and figure out an action plan. The workers decide that the shuttle service makes the most sense. It should solve the time and energy problem, and workers can keep their tools with them. To put the plan into effect, the project supervisor arranges for a large van and driver to provide the shuttle service.

Step 4: Test the solution to determine whether it actually works. The solution works, but there is a problem. All the workers are scheduled to start and leave at the same time, so there is not enough room in the van for all the workers and their equipment. To solve this problem, the supervisor schedules trips spaced 15 minutes apart. The supervisor also adjusts the workers' schedules to correspond with the trips. That way, all the workers will not try to get on the shuttle at the same time.

Step 5: Evaluate the process. This process gave both management and workers a chance to express an opinion and discuss the various solutions. Everyone feels pleased with the process and the solution.

Additional tips for problem solving

Know who is authorized to carry out solutions. Most companies appreciate employees who can solve problems creatively. However, you must solve problems within company rules and regulations. In the aforementioned example, although the workers came up with the solution, only the project supervisor was authorized to implement the solution by arranging for a van and driver.

Change negative thinking to positive thinking. When people get together to solve a problem, they tend to want to complain, which can waste a lot of time and sap everyone's energy and enthusiasm. To avoid this, get everyone to agree to withhold their complaints until the group has come up with at least a couple of solutions. Alternatively, you could give everyone on the team an opportunity at the beginning of the process to air just one grievance—to get it out of their system—and then move beyond any more complaining.

Learn to brainstorm. Many good ideas never become known because people are afraid they will sound silly. In a brainstorming session, you want to hear as many ideas as possible. You hear the phrase *out-of-the-box thinking* a lot these days. That means taking a fresh perspective on things. These solutions will not necessarily be entirely practical, but hearing different opinions, even those that seem off the wall, can help bring new ideas to the surface, which can then be used to come up with realistic solutions.

Don't settle for the cheapest, easiest, or fastest solution. It is important to save money, energy, and time, but a solution that is fast, cheap, and easy isn't necessarily a good one.

Don't complicate a simple problem. The best solution is not necessarily the most complicated one. Simplicity can be a virtue. It is not necessary to devise a complex solution when a simple one will do. For instance, say you are pretty high up on a scaffold and you need one of your tools you left on the ground. You just hooked in, and you don't want to unhook. So you call down to a co-worker and ask if there is some way to rig a contraption that could hoist the tool up to you. But that would mean the co-worker would have to look for rope and maybe a pulley, and take time out from a busy day to help you out, when the simplest solution might just be for you to unhook, go get the tool, and get back up on the scaffold.

What Would You Do?

You need to buy a new pickup truck to carry your tools to work. A neighbor down the street has one for sale, which your neighbor says "just needs a little work." You go look at it, and the truck barely runs at all. It's really inexpensive, though. Meanwhile, a local truck dealership is offering good deals on used but reliable pickups, but they cost more money. Do you buy your neighbor's truck, which you'll probably have to take to the mechanic numerous times to get it to run adequately, spending not only money but also a good deal of time? Or do you buy the more reliable one from the truck dealership, even though it is more expensive, and save yourself the hassle of dealing with frequent repairs?

Barriers to problem solving

Whether you are tackling a problem on your own or working with a group, you must be alert to obstacles to success. Watch out for the following hindrances to effective problem solving:

Close-mindedness. To find the best solution, you need to be open to new ideas. Sometimes the best solution is one you never considered. Remember, other people have good ideas, so you should be willing to listen to and support them.

Personality conflicts. Sometimes it's the person you least like on the team who offers the soundest solution. You may not get along with that person very well, but that shouldn't stop you from supporting that idea and trying your best to help implement it. Keep in mind that you recognize an idea to be the wisest choice even if you are not fond of the person who came up with it.

Fear of change. Many people are reluctant to change the traditional way of doing things. Most technological advances make work easier and more productive, so always be willing to learn new technologies and work with new equipment. In fact, as a construction professional, it's your duty to keep up with advances in your field.

Summary

To have a successful career in construction, it is important to develop sharp critical-thinking and problem-solving skills. When you pay close attention to the information you hear or read, when you evaluate that information guided by your own knowledge and experiences, and when you validate that information through further research and inquiry, you'll make sound decisions on the job and avoid jumping to ill-informed conclusions. Effective problem solving means carefully defining a problem, considering all the different ways it could be solved, choosing what you believe is the best solution, testing the solution to see how well it works, and evaluating the effectiveness of the entire process.

Here is a quick quiz that allows you to apply what you've learned in this module. Select the best possible answer.

1. When you think critically, you _____.

    a. pay close attention to information and evaluate it

    b. create a list of things that are being done incorrectly and then criticize them

    c. find fault with the work of others after carefully evaluating it

    d. think of a fast, cheap, and easy way to solve a problem

2. Good critical thinkers make _____ decisions.

    a. quick

    b. reasoned and informed

    c. flexible

    d. tentative

3. At a sales presentation for a new backhoe, the sales representative shows a video, hands out glossy brochures, and says that the backhoe works twice as fast as any other backhoe on the market. If you use your critical-thinking skills, you would _____.

    a. evaluate the product based on the information you get from the video and brochure

    b. evaluate the information based on the company's reputation and your knowledge of backhoes

    c. decide to buy the backhoe because the representative says it works twice as fast as any other

    d. decide not to purchase the backhoe because the salesperson seems too slick

4. You see an advertisement for a class that promises to prepare you for a certification exam you want to take. The ad says you'll be prepared for the exam once you take the class, which runs over three two-hour sessions. Experienced co-workers, on the other hand, say it takes at least 15 hours of study to prepare for the exam. What would be your best course of action regarding whether or not to take the class?

    a. Ask your boss to make the choice for you, since your supervisor passed the same exam years ago.

    b. Study for the certification on your own for 9 hours, then take the 6 hours of classes, which will add up to the 15 hours your co-workers say you need to be prepared.

    c. Decide not to take the class, since you never were much good with classwork to begin with.

    d. Find out more about the track record and reputation of the sponsor of the class and, if you can, ask the certification examiner what he or she thinks about the claims made about the class by its sponsor.

5. Among the things that improve the problem-solving process are _____.

    a. defining the problem and not trying to minimize or avoid it

    b. jack-in-the-box thinking and experimentation

    c. shutting off debate and choosing the solution based mostly on cost

    d. being skeptical of all solutions and waiting for the problem to resolve itself

6. Which of the following is an important step in problem solving?

    a. Getting input from those who are affected by the problem and from those who will be affected by any solution.

    b. Immediately discounting solutions that seem to be impractical.

    c. Creating as complex a solution as possible because the more complicated a solution, the better it usually is.

    d. Making sure co-workers you don't like are left out of the decision making.

7. When a solution does not work after you test it, you should _____.

    a. toss it out entirely and start the problem-solving process over

    b. think about whether the solution could be improved, or whether aspects of it could be adapted to another solution

    c. redefine the problem so that it will fit the solution

    d. tell the person primarily responsible for the solution to quit wasting everyone's time

8. You call a lumber company about a delivery, only to learn that the one scheduled for Monday will not occur until Wednesday. The best way to solve this problem would be to _____.

    a. tell the supplier to either make the scheduled Monday delivery—even if the supplier has to delay some other delivery—or your company will stop doing business with them

    b. inform your supervisor about the problem, and offer to go pick up the lumber yourself

    c. meet with your crew, ask them to brainstorm ideas to solve this problem, and then take the best one to the supervisor

    d. decide that it's a management problem and not your concern

9. Your company has used Brand X cement for awhile now because it's a good and reasonably priced product. Some workers have tried the less-expensive Brand Y cement and found it hard to work with. The manufacturer of Brand Y now claims to have developed a new and improved Brand Y cement that still costs less than Brand X. Your supervisor is thinking about changing to Brand Y. When your supervisor asks for your opinion, you say, _____

    a. "I'm not sure, boss. Whatever you decide is OK by me."

    b. "We tried Brand Y in the past and it was no good then. Why should we think it's any better now? Because the company that makes it says so? I think they're just putting lipstick on the pig, frankly."

    c. "We all make mistakes, and that's how we get better. The Brand Y company has learned from its mistakes, and so its cement has to be better than Brand X, whose company has never made a mistake and so has no inspiration to improve its product."

    d. "We could test Brand Y on a small project first, and see how it does. If it is easier to work with than before, we could then seriously consider changing brands. Do we know of anyone who is using the new Brand Y now? We could also ask their opinion."

10. Your co-worker T.J. has a track record for coming up with clever ideas. T.J. is also pretty arrogant, and both you and your co-workers find that attitude hard to take, even in small doses. Your boss, who thinks the world of T.J., has told you to include T.J. on the team you are putting together to solve a drainage problem. You should _____.

   a. explain to your boss that the other workers cannot stand T.J.'s attitude and that no matter how good T.J.'s ideas are, the team won't be worth much because of T.J.'s negative presence

   b. suggest that your boss just ask T.J. to solve the problem alone since T.J. is such a wonderful problem solver

   c. do as your boss asks, but freeze T.J. out of your team's discussions

   d. do as your boss asks and focus on getting good ideas from everyone, T.J. included

Individual Activities

Activity 1: Separating Fact from Opinion

In this activity, you will practice identifying facts and opinions, and separating one from another.

Example 1: Opinion vs. Fact

Consider the following statements. Notice that the two opinion statements combine both fact and opinion. Do you think these opinions misrepresent the fact that is presented?

Fact: In 2007, there were about 380,500 recorded cases of job-related injury and illness among workers in the construction industry, according to statistics from the Bureau of Labor Statistics.

Opinion: Construction work can be dangerous, so only people who aren't afraid of much should do it.

Opinion: Construction work can be hazardous, so you probably have to be a little crazy to do it.

Example 2: Wearing Hard Hats: Right or Wrong?

Read each of the following examples. In the spaces provided, write F after sentences that are facts and O after sentences that are opinions. Use the critical-thinking tools you have learned about in this module to carefully read and evaluate each statement.

The Occupational Safety and Health Administration (OSHA) says that wearing hard hats reduces workplace injuries. ___.

However, I think that construction workers should not have to wear hard hats if it makes them uncomfortable. ___.

I have talked to some of my co-workers and they agree with me. ___.

I hate wearing a hard hat. ___.

I once worked for a whole hour without my hard hat on, and except for the boss yelling, nothing happened to me. ___.

That goes to show you that wearing a hard hat is not necessary. ___.

Sure, a hard hat can protect your head if something falls on it. ___.

But that doesn't happen too often and it only happens to people who aren't paying attention. ___.

Since I always pay attention on the job, nothing will ever fall on me. ___.

After all, nothing ever has. ___.

I'm sure most people would agree with me that you shouldn't have to wear a hard hat if you don't want to. ___.

Work is hard enough without having the government telling us what to do. ___.

That's why I say that making us workers wear hard hats when none of us wants to is wrong. ___.

Example 3: Paint or Wallpaper?

My clients want me to hang wallpaper in their kitchen. ___.

Hanging wallpaper is a pain in the neck and painting is much easier. ___.

That's why I prefer painting to hanging wallpaper. ___.

Besides, paint looks better on a wall than wallpaper. ___.

I have been painting and papering for 10 years. ___.

My experience has been that, in kitchens, wallpaper generally does not hold up as well as paint. ___.

I will paper if the clients insist, but they will not be happy two years from now. ___.

Depending on the wallpaper they pick, the job will probably cost more if we paper. ___.

Activity 2: Evaluating Information

You receive the following letter. Use your critical-thinking tools to carefully read and evaluate this letter.

Congratulations! You are one of the lucky few especially selected to receive **this incredible offer**. We know how important **reliable construction equipment** is to you and your business. And we know that keeping the lid on **your construction budget** is also important. That's why we are making this **special offer** to you. Because of the downturn in the economy, we have a limited number of brand-new Best Backhoes available for purchase **for only $3,995.*** These backhoes are the finest backhoes available in the trade today. Just look at these great features:

- Tubular steel construction for added strength and longer life
- Wide support base for stable, safe operation
- Rubber pads for stable operation on hard surfaces included *at no extra charge*
- Locking pins for safety pivot points included *at no extra charge*
- Adjustable seat offering six-point adjustment (three more than the industry standard!)
- Flexible bucket operation from 9" to 36" and rotation from 165 degrees to 172 degrees

One look will convince you that the Best Backhoe is indeed the best backhoe you've ever seen. **But don't delay. Only 12 backhoes** will be sold at this budget-saving price **this weekend only**. Call 1-800-123-1234 for directions to our special sales site. Only you and a select few other customers will have this **once-in-a-lifetime opportunity**. Don't miss out on this great opportunity. This offer will never be repeated again.

Sincerely,

M. L. Smith
President, Best Backhoe Company

*Price based on Model 1XIJRM8. Prices for other models may vary. Taxes and additional fees not included. Does not include shipping. Purchasers may be required to provide advance credit information. No warranties, express or implied, apply to special sale items. All sales are final. No refunds. Best Backhoe Company is a member in good standing of the All States Construction Association.

Assuming you are in the market for a backhoe, here are some questions to consider:

1. How many people besides you do you think received this letter?

2. The letter explains why this deal is being offered. Why do you think the deal is being offered?

3. Do you know anything about Best Backhoes, the Best Backhoe Company, or the All States Construction Association? How can you find information about them?

4. How much does a brand-new backhoe actually cost? Is this price really a deal or is it too good to be true? How can you find out?

5. Are you getting a good deal because some features are included at no extra charge? How can you find out whether these items are standard on all backhoes?

6. Is an adjustable seat an important factor in choosing a backhoe? What is the industry standard for this feature?

7. What "advance credit information" do you think the company wants from you? Why do you think the company wants it?

8. What is the meaning of the sentence (in the smaller print) "No warranties, express or implied, apply to special sale items"? If you don't know, how can you find out?

9. Whom would you show this letter to for advice?

10. Based on your critical review of the letter, what will you do?

Activity 3: Evaluating Information

What follows are three construction problems, each of which has a problem-solving grid. Your goal is to find the best solution for these problems. Complete the grid for each situation. Follow the problem-solving process you learned in this module. (Note: An additional grid is also provided. Your instructor may give you an additional problem to solve.)

Problem 1: It's 7:00 a.m., and you are installing bathroom fixtures in a new house at a large subdivision. When installing the toilet, you realize that some parts are missing. Also, some of the parts included with the tank kit appear to be the wrong ones. Your co-worker has taken the truck and the cell phone to another site and won't be back until this afternoon. The nearest parts store is 10 miles away. You must finish this project today.

Five-Step Problem-Solving Process
Step 1: Define the problem.
Step 2: Think about ways to solve the problem.
Step 3: Pick the solution that seems best and figure out an action plan.
Step 4: Test the solution to determine whether it actually works.
Step 5: Evaluate the process.

Problem 2: Workers have to carry lumber from the back corner of the lot, where it is stored, to the working area. This wastes time and energy. There is no room to store the lumber close to the working area because the office trailers for the jobsite take up the space.

Five-Step Problem-Solving Process
Step 1: Define the problem.
Step 2: Think about ways to solve the problem.
Step 3: Pick the solution that seems best and figure out an action plan.
Step 4: Test the solution to determine whether it actually works.
Step 5: Evaluate the process.

Problem 3: *(your worksite problem)*

Five-Step Problem-Solving Process
Step 1: Define the problem.
Step 2: Think about ways to solve the problem.
Step 3: Pick the solution that seems best and figure out an action plan.
Step 4: Test the solution to determine whether it actually works.
Step 5: Evaluate the process.

TOOLS FOR SUCCESS: CRITICAL SKILLS FOR THE CONSTRUCTION INDUSTRY

Activity 4: Evaluating Alternatives

As we've seen, there is often more than one solution to a problem. How do you choose among several alternative solutions, particularly when all of them have both strong points and weak points? One way is to rank the solutions from the most workable to the least workable.

To complete this activity, read each case study and rank the solutions from 1 to 5 (see the following scale). Think in terms of which solution might be the best and which solution might be the most practical to apply. (You can also make this exercise a group activity by splitting into teams and then discussing your rankings with your teammates after each of you has read the case studies and ranked the solutions. Then your team can come together on a group decision on which solution is best.)

Ranking Scale

1 I think this solution will work the best; it's the most effective and the most practical to implement.

2 This solution is okay, but I don't feel it will work as well or be as easy to implement as my first choice.

3 This solution is also okay, but the first and second solutions seem to me to be easier to apply and more effective, and I'd want to try them first before applying this solution.

4 This solution may work, but I don't feel comfortable about it: it might not work and it might not be practical.

5 Even though this solution may work, I just don't like it. I don't think it will work, I don't think it's practical, and I think it might in fact lead to other problems.

Team Members (optional)

1. _____   2. _____

3. _____   4. _____

Case Study 1: Problem with a Co-Worker

You and E.D. have been working together a long time, and you've become good friends. You've gotten to know E.D.'s habits and behaviors pretty well. E.D. has always been reliable, but over the past two weeks, his performance seems to be slipping. You've noticed the following problems with him:

- Increased tardiness
- Bloodshot eyes
- Problems with balance when walking
- Making a lot of small mistakes

You've had to spend some of your own work time catching and fixing E.D.'s mistakes. Today he made a bigger mistake, and you were almost injured. You suspect that E.D. may be drinking before coming to work.

Rank these solutions (from 1 to 5):

A. _____ Take a wait-and-see approach. Give this situation a few more weeks to see whether the problems go away; meanwhile, continue to fix any mistakes made by E.D.

B. _____ Take E.D. out for a drink and ask what's going on. Be supportive but firm. Explain that you are worried about your own safety and about his well-being.

C. _____ Ask your supervisor to deal with this problem. Tell your boss you don't know how to bring up the subject, and you are afraid that you will wreck your friendship if you do.

D. _____ Be sympathetic to E.D., but understand that you have to watch out for your own safety. Ask your boss to assign you to a new partner and make up some excuse for your request so you don't have to tell on your friend.

E. _____ Write an anonymous note to E.D., listing the problems you've noticed. State your concerns about his safety and the safety of other co-workers. Explain that unless E.D. straightens out, someone will have to let the boss know about the problems.

Case Study 2: Working with a Difficult Co-Worker

You've been assigned to a project that requires a team of four workers. Your team members are:

- T.K., a friend of yours

- D.T., who does not talk very much but does a good job

- J.R., who argues with everyone and is difficult to work with

The job is scheduled to last eight weeks but after only two weeks, J.R.'s attitude is driving you crazy. You wish J.R. were not on your team, and there are still six weeks to run on the project.

Rank these solutions (from 1 to 5):

A. _____ Wait until after work one day to meet with J.R. Have a calm, professional discussion with him about his behavior and its effect on the project.

B. _____ Find out if other members of the team are having the same problem with J.R.'s attitude. If so, you can all talk to J.R. together.

C. _____ Just put up with J.R. for the duration of the project and ignore the stress this situation is causing you and your team.

D. _____ Give the situation another few weeks to improve. If things don't get better, you will ask your boss to talk to J.R.

E. _____ Complain to J.R. in front of the other team members. That way all the team members can talk about their feelings out in the open.

Activity 5: Brain Busters: Solving Puzzles

Solving puzzles is good exercise for your brain and can be fun as well. Try solving each of the puzzles below. (You can also make this exercise a group activity, by splitting into teams and comparing your solutions with those of your team members.)

Team Members (optional)

1. _____   2. _____

3. _____   4. _____

Puzzle 1: You want to finish the basement in your four-year-old home. You have to figure out how to get sheets of drywall down to the basement. The stairs from the first floor of your house to the basement turn at a 90-degree angle, so there's a landing halfway down. The builder installed drywall on the full length of the stairwell and painted it when the house was built. The basement windows measure 30½" × 24". A standard sheet of ¾" drywall at your local home improvement center measures 4' × 8'. There are no other ways to get into your basement. You don't want to cut down your drywall sheets because that will increase the amount of mudding and taping you will have to do. You figure out how to get the drywall into the basement without cutting it down. How?

Puzzle 2: You are assigned to drive covered truckloads of fill dirt onto a construction site. To get the most out of the deliveries, the fill-site supervisor makes sure that all the trucks are fully loaded and then some. The supervisor also insists that all loads be covered with a heavy-duty tarp so that no fill blows off during the trip to the construction site. Between the fill yard and the construction site are a series of overpasses. You drive under five of them with no problem. However, at the sixth overpass, you get stuck. What can you do to get moving again?

Puzzle 3: You have to cut a square piece of wood into eight equal pieces. However, you are only allowed to make three separate cuts. How will you do this?

Puzzle 4: The construction site for a large job employs workers in shifts 24/7. Because of security concerns, a 10-foot-high chain-link fence surrounds the site. A sudden heavy storm causes a flash flood in a nearby stream. When the storm ends, there is a deep, wide, muddy puddle between the parking lot and the gated front entrance to the construction site. The workers don't want to carry their heavy tools through the mess. One worker proposes building a temporary bridge but is told that is not allowed. Another worker proposes draining the puddle but is told that's not allowed either. Yet the workers manage to come up with a way to get from their cars to the site without getting wet or muddy. How did they do it?

Puzzle 5: You are given two empty containers: one three-gallon container and one five-gallon container. Neither container has any measurement markings on it. You are told to make only one trip to a large vat filled with paint and to bring back only seven gallons of paint. You must go directly to the vat, get the paint, and return directly to your workstation. You don't have any way to measure and mark the cans, yet you successfully bring back exactly seven gallons of paint. How?

Activity 6: Using Logic to Sort Out Possible Solutions

Earlier in this module, you learned how workers solved the problem of getting from a distant parking lot with tools and equipment to a jobsite. Before those workers chose a shuttle service to solve their problem, they had to consider four possible solutions. In this example, you will learn how they used logic to consider each option.

Here are the solutions the workers considered:

Solution 1: Install lockers for tools and equipment closer to the worksite.

Solution 2: Have workers drive up to the worksite to drop off their tools and equipment before parking.

Solution 3: Bring in another construction trailer where workers can store their tools and equipment for the duration of the project.

Solution 4: Provide a shuttle service from the parking lot to the worksite.

Here is how the workers logically considered each possible solution:

Step 1: Will the proposed solution cause other problems?

Yes. Choose another solution and ask this question about that solution.

No. Go to Step 2.

Step 2: Can we come up with a workable plan to carry out this solution?

Yes. Go to Step 3.

No. Choose another solution and return to Step 1.

Step 3: Does everyone agree that this solution meets our needs?

Yes. Choose this solution.

No. Choose another solution and return to Step 1.

In a similar manner, use logic to consider each option in the following scenario.

Team Members

1. _____   2. _____

3. _____   4. _____

Assume that you and your team members have to select only one meal for the entire group to eat for lunch. Assume also that you have collected $30 to pay for the group's lunch. Here are your options:

Option 1: Steak sandwich and fries: $7.50 per person

Option 2: Large pizza with pepperoni and sausage: $15.95

Option 3: Pulled pork sandwich and coleslaw: $6.25 per person

Option 4: Beef and bean burrito platter with salad, guacamole, and sour cream: $5.45 per person

Now answer the following questions, in this order, for each option:

Step 1: Will choosing this option cause other problems? (For example, some workers may have an allergy to certain foods.)

If you answer yes, then list the problems and keep them in mind when you ask this question about another option.

If you answer no, go to Step 2.

Step 2: Will our budget cover this option?

If you answer yes, go to Step 3.

If you answer no, you can do two things: choose another option and return to step 1, or modify the budget ($30) and proceed to Step 3.

Step 3: Does everyone agree that this is the best choice?

If you answer yes, then go enjoy lunch.

If you answer no, choose another option and return to Step 1.

Your team may decide that it does not like any of the options presented here. If so, and if time allows, come up with other options by asking each team member to contribute one. Then use the logical process to choose one of those options.

Resolving Conflict

"…there was never a time during my command when I would not have chosen some settlement by reason rather than the sword."

— Ulysses S. Grant,
Civil War general
and U.S. President

Introduction

People don't always get along. Disagreements and conflicts are inevitable. The trick is to prevent smaller disagreements from blowing up into major conflicts. A conflict, even if it is sparked by a seemingly minor issue, can grow into a feud that can make the entire workplace unpleasant and workers' jobs and lives unnecessarily difficult. Conflicts that are not resolved can lead to long-term resentments that put schedules and projects at risk. It's important, therefore, that you know how to resolve conflicts quickly, reasonably, and rationally.

If you work hard and well, treat others with respect, and do your best as part of the team, you should be able to develop and sustain good relationships with most of your co-workers and supervisors during your construction career. Sometimes, however, you'll get into arguments that can strain those relationships. In this module, we offer tips and techniques for avoiding conflict and for quickly and effectively resolving those that do occur. This module also includes some suggestions for dealing with difficult co-workers, so that you can avoid conflict with those who like to stir things up.

The causes of conflict

Disagreements can arise from just about anything, from differences in opinion to differences in personality and age. Disagreements can lead to conflict, and since disagreements are common, so are conflicts. Sometimes the conflict lasts only an hour or so; sometimes the conflict, left unaddressed, can go on much longer. Some people avoid conflict as much as possible; some people act as if it doesn't exist; and some people seem to like it and even look for it. Not all disagreements lead to conflict, not all conflict spirals out of control. Not all conflict is necessarily bad or even unpleasant. When we have the skills to handle conflict, we can manage conflict to learn more about each other and come up with new ideas.

It's vital that you know how to manage conflict on the job. The first step is to know the common sources of conflict, so that you can try to prevent the conflict in the first place.

Conflict with co-workers

During your construction career, you are likely to be around your co-workers as much as (and sometimes even more than) your family. There are many reasons why co-workers experience conflict. The following are five of the most common causes of conflict at work. Each cause listed includes an example of how it might spark a conflict on the job, along with two suggestions—one good, one not so good—on how to handle that particular situation.

Different work habits

People have different habits, patterns, and tempos of working. Some are neat; others may seem messy and disorganized. Neat people might think messy people are too disorganized to get anything done; messy people might think neat people waste too much time tidying up. Some people work fast, others more deliberately. Faster workers may think more methodical workers are just slow and unproductive; more deliberate workers may think faster workers are careless.

Example: You like to organize your tools and think through tasks before you start working. Your co-worker likes to dive right in, get in a groove, and get the job done. You are both equally productive.

Poor approach: You tell your co-worker, "You're driving me nuts the way your rush around and leave stuff everywhere! You're going to make a bunch of mistakes if you don't slow down!"

Better approach: You think, "My co-worker rushes around a lot, but that doesn't influence how I do my job. And since we both get the job done and done well, why should it matter that my co-worker and I work at different paces?"

Different attitudes toward the company or job

Some people love their jobs and jump out of bed in the morning, ready to take on the world. Other people hate getting up in the morning and work just to pay their bills. Some people get along well with their supervisors; others don't.

Example: Your co-worker T.J. complains about everything from the taste of the morning coffee to the way the boss handles things.

Poor approach: You listen to this co-worker and think, "T.J. is a pain in the neck, with all that complaining. T.J. should just quit, and maybe I'll tell T.J. that one of these days."

Better approach: You tune out your co-worker's constant complaining and think, "I like this job. It's unfortunate T.J. doesn't, but I shouldn't let that influence me. Whether T.J. likes the job or not doesn't influence me one way or another."

Differences in personality and appearance

People act and express themselves in widely different ways. Some people are quiet and keep to themselves; others are talkative and outgoing. A quiet worker may get irritated by a chatty one; a gregarious worker may think someone who is reserved is stuck-up. Differences in hair or clothing styles or even in jewelry can cause conflict among workers. You don't have to necessarily admire how your co-workers look, dress, or express their personalities, but you can appreciate how they work. Unless your co-workers' appearance or personality is a detriment to the job, the work environment, or workplace safety, they're as entitled to their preferences as you are to yours: remember, it's important to respect your teammates. If someone is teasing you about your appearance or personality, ask yourself whether it is worth getting bothered by it and letting it lead to an argument.

Example: Your co-worker makes fun of your new haircut.

Poor approach: Make fun about some aspect of your co-worker's appearance in response; next thing you know, you're trading insults, and the conflict is intensifying and will be that much harder to defuse.

Better approach: Ignore the teasing. Your co-worker is being childish. If you choose to respond with an insult, you'll be lowering yourself to your co-worker's level and aggravating the situation, and then who knows how long it will go on?

Differences in age

People born in different generations may see the world differently. Older workers might think their younger counterparts lack a good work ethic. Younger workers might think older workers are stuffy and old-fashioned. Both may think they have nothing to learn from the other, and both would be wrong.

Example: You are assigned to work with someone from a different generation who does not believe in doing things the way you do.

Poor approach: You think, "This person just doesn't want to do things my way because this person doesn't respect people of my age. I wish I could work with someone my own age. This person is just going to slow me down."

Better approach: You think, "This could be a win–win situation here; there might be something of an age or generation gap between my co-worker and me, but if we're both open-minded we probably have a lot to learn from each other because we're different ages. So I'm going to make the effort to bridge the gap and listen respectfully to my co-worker's suggestions, while offering some of my own."

Problems outside the job

Stress from personal problems at home can lead to stress at work, and that stress can spark conflict or aggravate disagreements so that they become quarrels. Try to keep personal problems away from work, and know when it's OK to ask your supervisor or co-workers for help or understanding. Likewise, if a co-worker has personal problems and that's spilling over into increasing conflict with you, consider whether your co-worker needs some understanding or space.

Example: Recently, you've had several serious family-related problems; you're arguing frequently with your spouse, and you and your sister disagree strongly about what to do with your ailing mother. Therefore, you're having some problems focusing on work right now.

Poor approach: Don't tell anyone at work about your problems or concerns because it won't do any good. If anyone hassles you, warn that person you are not in the mood; if anyone asks you if something's wrong, tell that person to butt out and that it's your business. And if your boss tells you once more to keep your mind on your job, tell your boss to quit bothering you because you have problems and the boss isn't going to make your work any better by giving you a hard time.

Better approach: Talk to your boss and co-workers. Tell them in general terms that you are having a rough time. Say that you appreciate their patience and understanding while you work through your problems. Most supervisors and co-workers will understand, and some of them might sympathize and offer to help. If a co-worker isn't satisfied with your level of explanation, however, and starts probing to find out more detail about your situation—more than you were willing to share—then you might not want to talk about your situation much more with that co-worker. That person might not have your best interests at heart and instead just wants to find out more details to fuel the rumor mill.

When you are new on a job, you may occasionally encounter people who feel threatened by new employees for one reason or another. If you are new to a crew and you think you are making your co-workers uncomfortable, let them know that you are glad to be on the team and that you hope you can learn from their experience.

Conflict with supervisors

Most workers experience conflict with their supervisor at some time. Such a situation can be unpleasant and potentially worrisome. Here are three common reasons for conflicts with a supervisor; with each reason, there are two possible approaches (again, a poor one and a better one) for handling the situation.

Workload

A supervisor's job is to keep workers as productive as possible. Workers can become resentful, however, if they feel they're being pushed too hard or given too much to do in too short a time, especially if they're already doing a lot of work.

Example: Your boss announces that overtime will be required to meet a scheduled deadline, even though the work crew has been putting in a lot of volunteer overtime already.

Poor approach: You think, "I am sick and tired of this. It's not my fault the job is behind schedule. I did my part and volunteered, and now they're making me come in after hours! The heck with it, I'm leaving at the regular quitting time. I dare my boss to fire me, after all the hard work I've done recently."

Better approach: You think, "Weather's been bad, and the boss can't control that. Many people have been absent, too, since there's some kind of flu going around. The boss can't control that, either. The boss needs us to catch up to the schedule and since I'm a professional, I'm going to do my part and do the OT. After all, my boss is putting in a lot of overtime, too, and is under a lot of strain."

Absenteeism and lateness

Both of these can lead to serious delays, and therefore both can lead to significant conflict with your boss. By arriving on time for work, calling in sick only when you're legitimately ill, and always notifying your boss before your shift starts if you're going to be late or absent, you'll steer clear of a major source of conflict between bosses and workers.

Example: You miss your bus and have to wait for the next one, which is scheduled to arrive in 30 minutes.

Poor approach: Relax and read the paper until the bus comes.

Better approach: Do whatever you can to contact the boss. If you have a cell phone, call the boss right away. If you are unable to contact the boss, you must use this as a learning experience so that you will never again be caught unable to communicate with your boss when you are running late. If you're not on the job when you're supposed to be and no one knows why, that absence will probably upset your team's work schedule and your boss's plans for that day; it could even damage the overall project schedule. If you do not contact your boss when you are running late, no matter what the excuse, this could have serious consequences. Unexplained absences always have a significant impact on the project schedule and cost.

Criticism

Don't take criticism personally; everyone is on the receiving end at one time or another. It's a valuable learning tool, and it's usually constructive. If you believe the criticism is not appropriate, however, or that the criticism has been delivered too harshly, you should discuss the issue with your supervisor, calmly and clearly.

Example: Your boss says that you are forgetting to do certain tasks and you need to start managing your time better. He recommends that you write down your work assignments each day so that you'll know to leave time for each.

Poor approach: You snap at the boss, "Why are you always on my case? I guess nothing I do is good enough around here. You never seem to nag others like you do me. Why can't you just leave me alone? I'm doing the best I can!"

Better approach: You think, "I don't like getting criticism, but that's part of what my boss is supposed to do. Come to think of it, my boss is right: I have forgotten to do a few things recently. I'll try this advice and write down all that I need to do. After all, the boss is just trying to help me to do a better job; it's nothing that I should take personally."

Simple ways to prevent conflict

Experienced supervisors try to build teams of people who work well together and get the job done. Even in the most cohesive teams, however, conflicts can still occur, many of them caused by people getting irked over insignificant matters. You can do your part to help avoid turning minor irritants into conflicts by thinking instead of reacting if someone is cruel or insulting toward you; realizing that most people are not trying to deliberately annoy you; and staying out of arguments between others.

Don't react; take a moment to think

If someone is nasty or insulting to you out of the blue, don't snap back with the first nasty insult you can think of in response. That will just escalate the situation. Instead, remain calm and take a deep breath. Try not to get angry. If you already are, manage your anger as quickly as possible. Walk away to collect your thoughts, if necessary. You don't have to prove yourself by responding with another insult. What would it prove—that you can trade insults? Is that going to get the job done? Will that lead to a promotion? If you are tempted to fire back with some choice words, ask yourself first, "Is this going to matter a week from now?" You will realize that in all likelihood, it will not; it is much more sensible to ignore the situation. If someone snaps at you, ask yourself whether it might have been prompted by something other than the person's bad attitude. Think whether you might have done something that irritated the co-worker.

Example: Your co-worker calls you a lazy, worthless pig—the same co-worker from whom you borrowed a saw without asking last week.

Poor approach: You say, "I'm a pig? You're the messiest person around here. When you leave your work area, you leave materials all over the place and don't bother to clean up; it's a wonder people don't fall over your mess and break a leg."

Better approach: You think, "OK. This person's upset because I took the saw without asking. I'd probably be irritated too. I should have asked to borrow the saw." Then defuse the situation by saying, "If this is about the saw I borrowed last week, I shouldn't have taken it without asking. I apologize; I won't do it again. Let's not quarrel over it and start calling each other names."

Don't take it personally

You may sometimes think that other people do things or act in certain ways simply to annoy you. But that is rarely the case with most people. Recognize that others developed their habits and personalities long before they met you.

Example: Your co-worker, T.J., has a laugh that sounds like a jackhammer. You hear the real thing enough during the day, and T.J.'s laugh is really starting to get under your skin.

Poor approach: You think, "T.J.'s laugh is driving me nuts. I wonder if T.J. is doing it deliberately to annoy me, knowing that it bothers me? Next time I hear it, I'm gonna lose it, I swear."

Better approach: You think, "Maybe T.J. can't help laughing like that. I notice it gets worse when T.J. is really stressed. It probably has something to do with anxiety. I know I can show stress in unusual ways myself."

Avoid getting involved in other people's arguments

You may feel that you can help other people resolve their conflicts. Generally, however, it is better to stay out of the situation. Even if one of the people involved is your good friend and the other is someone you are not too fond of, you should mind your own business. If you suddenly find your name brought up during the argument, try to avoid getting involved; but if you do get involved, do not take sides.

Example: Two co-workers get into a fight over who made a mistake. One of them is your good friend J.K. The other is D.D., someone you've had issues with in the past. J.K. turns to you and says, "You saw it, right? D.D. screwed up, not me! You've had the same problems with D.D. as I do now!"

Poor approach: You say, "You're right, J.K., it's not your fault. D.D., why are you always messing up?"

Better approach: You say, "All I see is there's a problem here, and it needs to be fixed. I don't care who's at fault, J.K.; let's all figure out a way to fix this, and not waste time pointing fingers."

Admitting When You're Wrong

Apologizing to a co-worker when you are the one at fault is not a sign of weakness but of strength. It shows you are adult enough to take responsibility for something you have done wrong. Often, it is hard to determine who is at fault; sometimes no one is, sometimes everyone is. You are not under any obligation to apologize in those cases. If you are in a dispute with a co-worker and you are the one who is clearly at fault, however, you need to apologize so that the dispute can be put to rest. For instance, if your conflict with a co-worker started about the time you took one of that co-worker's tools without asking and then broke it, then it is your responsibility to apologize. Make your apology short and sincere, then get back to work.

Resolving conflicts quickly and effectively

The methods we've been discussing can help prevent minor disputes from spiraling into major conflicts. But there are conflicts that aren't as easy to resolve, or issues that you just cannot walk away from or ignore.

These types of conflicts need to be resolved as quickly as possible. An ongoing conflict between co-workers can turn into a feud, which will affect the morale of everyone on the jobsite, and the longer this feud goes on, the more people who weren't initially involved tend to get involved. For the good of the project, as well as for everyone's mental health, such serious conflicts must be dealt with quickly. In this section, you will learn how to resolve such conflicts, whether they are with your co-workers or your supervisor.

Resolving conflicts with co-workers

Conflicts with co-workers, especially if they are allowed to continue, can be a real problem at the jobsite and a detriment to the project. Presented below is a five-step process for resolving conflicts; for this process to work, however, all parties to the conflict must first agree on some ground rules:

- Put emotions, especially anger and defensiveness, to one side.

- Agree to be polite and respectful of one another.

- Keep an open mind.

- Listen to others without interruption.

Step 1: **Bring the conflict into the open.** Everyone must first admit a conflict exists; otherwise, the rest of the process will not work.

Step 2: **Discuss and analyze the reasons for the conflict.** Everyone involved should try to discover what caused the conflict in the first place and what has made it worse since then. This doesn't mean insisting that a particular person started the conflict or that a particular person wouldn't let the matter drop. Be specific when discussing what might have caused the conflict and what might be perpetuating it; for instance, "The reason I think there's a conflict is because so-and-so cursed me the other day when I asked for help on a project, even though I asked nicely and the boss told me to."

Step 3: **Develop possible solutions.** You do this in two ways: through collaboration, where all parties to the conflict work together to find a solution acceptable to everyone; and through compromise, where each party to the conflict gives up something so that everyone involved gets something out of the resolution.

Step 4: **Choose and carry out a solution.** Everyone involved must agree on a solution and make a commitment to do things that will make the solution work.

Step 5: **Evaluate the solution.** Revisit the problem to clear up unresolved issues.

Collaboration and compromise

Collaboration and compromise are skills that take time and practice to develop. There are no specific steps to collaborating and compromising; you just have to feel your way (along with the other party in the dispute) as you go along, depending on what the conflict is. There are some things you can do, however, that will help:

- Show respect for others at all times.

- Listen at least as much as you talk.

- Search for common ground.

- Be prepared to give up a little of what you want.

- Admit when you're wrong, and apologize for your mistakes.

- Accept that other people think differently than you do and have needs that differ from yours.

- Focus on problems and behavior, not on personalities and appearances.

- Talk about only one problem at a time; otherwise, you are setting yourself up for failure.

- Don't insult others or use profane language.

- Don't threaten others with words or actions.

- Avoid saying "you never" or "you always"; these phrases produce negative reactions.

- Keep any promises you make.

Resolving conflicts with supervisors

When you are resolving a conflict with a co-worker, you are dealing with someone who is your equal in the workplace. Your co-workers do not have authority over your work assignments. They did not hire you, they cannot fire you, and usually they are not held directly accountable for your work. When resolving conflicts with your boss, however, you need to take a somewhat different approach, because your boss *does* have authority over you. Here are some good practices to follow when resolving a conflict with your boss:

- **Gather your thoughts.** Think about the conflict and what caused it. It may help to write down your thoughts, which might put the situation in perspective. You may decide there's no real reason for the conflict, or you may come up with a simple solution to it. Think about what you are going to say before you meet with your boss, so that you do not waste time getting to the point.

- **Respect your supervisor's work schedule.** Don't march up to your boss and launch into your grievances. Instead, choose a time when your boss is less busy to talk about the matter. Some conflicts require more than couple of minutes to discuss; in those cases, you should ask your boss to set aside some time when it is convenient to meet with you. Let your boss choose the time. Ask to speak to your supervisor in private so that you will not be interrupted.

- **Stay calm and focused.** Speak slowly and carefully, and don't raise your voice. Focus on the facts; do not say things you cannot prove. Do not make accusations or threats, and don't be argumentative. Remember, your body language conveys as much as your words do. If your boss has given you 15 minutes to make your case, be sure to stick to that schedule; do not ramble off onto other topics.

- **Make your case clearly, and offer suggestions for resolving conflicts.** Keep your presentation of the case limited to the conflict between you and your supervisor, and explain how it is causing you problems and making you worry. Do not bring other issues into the discussion, such as the opinions or concerns of your co-workers. When offering suggestions for resolving the conflict, focus on things that will benefit the project or the team and not just you.

- **Respect your supervisor's decisions.** Once your boss makes a decision, it is final. Sometimes supervisors feel backed into a corner when their employees confront them. They may react hastily, then they'll think about the issue and make amends the next day. If your boss gets heated, give your supervisor time to make amends and then accept them gracefully. Remember, the idea is to resolve the conflict.

The following table presents two approaches, a poor one and a better one, to handling an example of a conflict common to a jobsite: a disagreement with your supervisor about overtime. In the better approach, these practices help in managing the dispute better; in the poor approach, the dispute is made worse by doing the opposite of these practices.

Situation: The boss has told you to work overtime this evening and says, "No 'if's, and's, or but's' about it." You came in late that morning without calling in, and your boss reprimanded you about it, saying such tardiness, especially without notification, is putting the project schedule at risk. Earlier, you were talking with your co-workers, who were complaining about how the boss is too picky about quality control, which is the reason why the project is behind schedule and everyone has to put in more overtime. You hadn't thought about that, and it makes sense to you. What's particularly frustrating is the timing: if you have to work overtime this evening, you'll miss a long-anticipated family event that's been in the works for some time.

Practice	Employing the Practice: The Better Approach	Doing the Opposite: The Poor Approach
Gather Your Thoughts	You analyze the situation calmly. You know that coming in late causes problems and the boss was therefore justified in reprimanding you; the reason you have to work overtime is probably related to coming in late. In planning what you are going to say, you focus on the core of the conflict: your boss wants you to work overtime this evening, but then you'll miss the family picnic you've been looking forward to so much. You think about whether there's a simple solution to the issue.	You don't think about what you're going to say before you talk to your boss, and you don't think of any solutions; you just brood about how unfair it is that you have to work overtime and how inconsiderate your boss is in making you miss a fun family event.
Respect Your Supervisor's Work Schedule	You pick the right time to talk to your boss, when he isn't busy. If your boss is busy, you ask when's a good time to meet. If you need more than a few minutes, you ask the boss to schedule some time at the end of the day to have your discussion. You talk about the issue in private.	Instead of respecting your boss's schedule, you march right into the office when the boss is talking to some co-workers, interrupt the conversation, and in front of everyone start listing your grievances and complaining.
Stay Calm and Focused	When you go in to talk to your boss, you stay calm and on topic. You apologize for being late and acknowledge how irresponsible it was not to call in first. You stress that you want to make things right and that you understand you need to work the overtime. You make your point, you offer a solution if you have one. You look the boss in the eye during the discussion, but you don't argue or act aggressive or put upon; you conduct yourself as a professional. You don't bring up what you heard from your co-workers about the boss and quality control; that's not the issue and not your concern.	You don't stick to the facts calmly; instead, you shout and argue. You make accusations instead of offering solutions. You tell the boss what the others are saying regarding quality control and how that is the reason why the project is off schedule and why you're being made to work overtime.
Make Your Case Clearly; Offer Suggestions	You stress what, in your opinion, is the root of the problem: that if you work overtime this evening you'll miss your family's picnic, which you've been looking forward to for a long time. Then you offer your solution: that you will work the overtime whenever else the boss needs you to—some other day this week or over the weekend—but that, if possible, you'd like to have this evening free.	You start shouting, "Why am I always the one who has to work overtime! You're always giving me trouble; you know there was a lot of traffic this morning, that's the reason I was late, it's not my fault, no wonder I have to work OT, I'll tell you what the fault is, you're such a control freak, we can't stick to the schedule, and you should hear what the others have to say about it. So go ahead and make me work overtime, I'll miss the family picnic I've been looking forward to for months, they always say family is the most important thing, but I guess your precious schedule and pickiness is more important…" You go on and on, making the situation worse and not better.

Practice	Employing the Practice: The Better Approach	Doing the Opposite: The Poor Approach
Respect the Final Decision	Whatever the boss's decision, you respect it and abide by it. If you have to work the overtime this evening in spite of your request, you realize you wouldn't have gotten your way by yelling and screaming; you might have been fired instead. You realize that by not agreeing to your question, your boss might be quite dissatisfied with your job performance, and you resolve to act more professionally in the future. You consider whether your boss's decision might have been different if you had let your supervisor know you were running late; if you had acted more responsibly then, your boss might have been more willing to let you work the OT some other time.	You've backed your boss into a corner by insulting, confronting, and accusing your supervisor in front of others. You have ensured that the conflict with your boss is going to get worse, and that is not good for you. If your boss wasn't willing to let you off from working overtime this evening, your behavior now could well be endangering your job. In the future—if you've managed to keep your job—you ask your boss for some favor, your supervisor is going to remember this incident. How do you think that will affect your chances?

General tips: Coping with annoying behavior

People usually aren't trying to be annoying. They're usually unaware that how they eat or talk or act may be affecting someone else negatively. It is important to remember that if you find your co-workers' behavior annoying, in all likelihood they aren't deliberately trying to annoy you. So give them the benefit of the doubt. If you are really bothered by something, you are much more likely to get satisfaction if you bring up the matter gently, and with a little patience and class.

Be tactful and consider the other person's feelings

Don't bring up an annoying behavior when other people are around. It's better to have a private conversation. Don't say things that could insult the person or result in hurt feelings.

Example: Your co-worker's sloppy eating habits are disgusting to you and others.

Poor approach: During lunch, say loudly, "You're disgusting! You eat like a pig! Shut your mouth when you eat!"

Better approach: In private say, "You know, we all enjoy those stories you tell us at lunch; no one tells them any better. It's just that I'd enjoy it more if you didn't talk with your mouth full."

Use words that will not make others defensive

Using the word *you* when talking to others about a problem can put them on the defensive. When people are on the defensive, it is harder to talk to them. When possible, try to use *I* or *we*. Using *we* emphasizes the importance of the team and teamwork.

Example: Your co-worker sometimes has body odor.

Poor approach: Say in front of everyone, "Don't you ever take a shower? I swear you smell so bad, the skunks run when you walk by."

Better approach: When a few team members, including the one who smells, are around and you're all sweaty on a hot day, you joke, "Look at us—the way we smell after a hot day, they ought to come up with an OSHA regulation. I probably stink more than any of you. Why don't we try a double dose of deodorant before we come to work every day?"

Use humor when appropriate

Often, you can let someone know about annoying behavior by taking a humorous or light-hearted approach. Be careful if you choose to do this, however; it works only when everyone is laughing along together and not if you are laughing at someone. Laugh at yourself for bringing the subject up, and not at the subject's behavior.

Example: Your co-worker has a habit of singing loudly while working, making it hard for others to concentrate on their work.

Poor approach: You and your co-workers groan loudly and clutch your ears every time the singing begins.

Better approach: Whenever the co-worker starts singing, you all join in, singing as off key as you can. Everyone together makes a real racket, and it sounds horrible yet hilarious. The co-worker whose singing started it all may laugh harder than anyone else. You all have a good laugh and then get back to work, and the singing employee gets the point.

Help the other person save face

In most cases, people are surprised and even embarrassed to learn that their behavior annoys you. Whenever possible, give others the chance to change their behavior without calling a lot of attention to it. Let people off the hook gracefully; they'll appreciate it.

Example: You notice that a co-worker has suddenly developed an offensive habit, _____.

Poor approach: In public say loudly, "Since when did you think you could start doing that around here? We're not animals in the zoo, you know? Well, maybe *you* are."

Better approach: In private say, "I don't want to embarrass you, but did you notice you've started _____ around here? That's not like you. Is there something wrong?"

Ask your boss for help, but only as a last resort

As a professional, you should be able to deal sensibly with behavior you find annoying in others. Most often, these incidents are minor matters, and you should be able to settle them with a minimum of fuss. In some cases, however, you may feel that the behavior is too serious to handle on your own. In these situations, you should ask your supervisor for help.

Example: Your co-worker is harassing you. You keep finding offensive notes at your work area. In addition, this person often pushes or grabs you in a threatening way and does things that embarrass and upset you. You have confronted your co-worker and demanded to be treated with respect, but the behavior just gets worse.

Poor approach: Ignore the co-worker's harassing behavior in the hope that it will eventually stop.

Better approach: Talk to your boss about this aggressive behavior. Provide details of the things your co-worker has done, and copies of the offensive notes you have received.

When in Doubt, Check It Out... in the Employee Handbook

Your company—whether it is a construction firm or a temporary agency that assigns you to companies—will have policies on what it considers acceptable and unacceptable behavior among its workers. You should become familiar with those guidelines and regulations. These policies can be found in the employee handbook. They cover the expected rules of conduct as well as when a company can discipline—that is, suspend or fire—a worker who does not follow these rules. These rules can cover such topics as:

- Alcohol and substance abuse, as well as policies for substance abuse testing

- Attendance

- Harassment, sexual or otherwise

- Guidelines on the acceptable use of e-mail, the telephone, the Internet, and company equipment and vehicles

If you feel you have been disciplined or fired unfairly, look in the employee handbook. Under what circumstances can an employee file a grievance about it? If you feel you must file a grievance about your boss, you should do it only as a last resort after you have tried your best to settle the matter. If you still cannot resolve the issue, you cannot just go talk to your boss's supervisor about it; that will get you nowhere and just makes matters worse. You must follow specified procedures for filing a grievance, which should be in the employee handbook.

The employee handbook will also have information on other topics such as pay, benefits, and policies for promotion.

Coping with difficult co-workers

Some people exhibit a specific annoying characteristic in much of what they do. They can make a person difficult to handle or be around. Some of these characteristics are listed below:

- **Blamers** refuse to take responsibility for their mistakes. They'll blame anyone who happens to be around instead of admitting that they sometimes make mistakes. To cope with a blamer, point out that blaming accomplishes nothing except to undermine the team and make it harder to get the job done.

- **Braggarts** and **know-it-alls** are always boasting—about their intelligence, their looks, their toughness, and their abilities. They've got the best cars and the most money in the bank. They do the best job and they have all the answers. When braggarts start blowing their horns, agree with them; tell them how much better they are than anyone else. That will catch them off guard and they'll be so shocked to have someone actually agree with them, they'll be speechless—which is something a braggart rarely is.

- **Brown-nosers** are always trying to get in good with the boss, through flattery or constant agreement. Sometimes these people are trying to separate themselves from the rest of the crew because they think they're superior. Sometimes, however, a brown-noser is just desperate; they may really need money, and feel obliged to act that way. If that's the case, try to empathize.

- **Complainers** never find anything that is good enough. If there is nothing to complain about, they'll find something. Complainers aren't interested in solutions. If a complainer is getting under your skin, tell them that finding fault never solved anything.

- **Delegators** are the ones who are always talking about the need to delegate their work, when what they really want to do is avoid having to do it themselves. Don't fall for this trick, even if the delegator's request is polite and reasonable. Tell them you have your own work to do, and that everyone on the team has to do their share.

- **Detailers** can take forever to finish a task. It's one thing to be thorough and careful; it's another thing to obsess over every little thing. If a detailer's slowness is affecting your job, tell the person that while you appreciate their thoroughness, all that fine-tuning is upsetting the work schedule.

- **Dictators** are, in essence, bullies. They harass and intimidate others to get their way, and they don't care about others' feelings. They make demands and give nothing in return. As with all bullies, the best way to handle jobsite dictators is to stand up to them. Be firm, but don't get emotional or angry.

- **Gossips** spread rumors. They are often unpleasant people who try to pit teammates against each other, and attempt to manipulate others for one reason or another. Don't listen to gossips, and don't tell them anything you don't want to be generally known at the jobsite.

- **Hotheads** lose their temper easily. Often they'll be sorry afterwards and even apologize, but the next thing you know, they're flying off the handle again. If you don't like dealing with people who lose their temper, it's best to ignore them and focus on your work.

- **Mood-swingers** are unpredictable; you never know what to expect from them. Some days they're happy as can be; other days, they're miserable. Avoid saying or doing things that might cause these people to react emotionally. Sometimes moodiness can be a sign of depression or mental illness; mood-swingers may not always be able to control their moods.

- **Snobs** expect to be treated better than everyone else. They are similar to braggarts in that way, except that snobs may not be quite as stuck on themselves; they are just less fond of others. If a snob is bothering you, point out that no one gets special treatment on the jobsite.

- **Tattletales** are similar to brown-nosers, but they take it a step further. Tattletales try to get in good with the boss by bad-mouthing others to look good in comparison. Be particularly careful what you say and do around tattletales.

Summary

Conflict may be inevitable, but it can be readily managed. Know when you should handle the situation yourself, when you might want to involve your boss, and when you should just let the matter drop. Many common causes of conflict with co-workers can be avoided. Don't react or get too emotional during a dispute, and stay out of arguments if you can. If you have a dispute with your supervisor, discuss it at a time convenient for them. Make your case calmly and respectfully and don't yell or make accusations; stick to the facts and accept your boss's decision. Always look for solutions to conflict, whether with your co-workers or your boss. Remember that most behaviors you may find annoying aren't done on purpose. You will come across a difficult co-worker on occasion—a bully, braggart, or a gossip, for instance—so it will be to your advantage to know how to interact with difficult people.

Here is a quick quiz that will allow you to apply what you have learned in this module. Select the best possible answer, given what you've learned.

1. Your co-worker often wears an "I Love the Yankees" T-shirt to work. You're a die-hard Yankees hater. You joke back and forth about it, but sometimes the joking gets a bit heated, and there's some tension between you two now. One day you and the co-worker, both of you not looking where you're going, run into each other, knocking each other down. You should _____.

   a. yell something obscene about the Yankees, figuring it's better to yell that than something obscene about the co-worker personally

   b. ask if the co-worker is all right and apologize for being so careless

   c. joke, "You run around here like the Yankees run the bases—without a clue," even though you are pretty angry with your co-worker for knocking you down

   d. grunt some kind of half-hearted apology and then say, "You know, you could make it up to me by taking me to a Yankees game"

2. Your boss calls you aside and talks to you about a serious mistake you made the day before. Your boss calmly explains that you will have to work overtime to fix the mistake. Later, a co-worker asks what happened. The best thing to say is, _____.

   a. "Mind your own business, it's between me and the boss and none of your concern"

   b. "Eh, I kind of butchered a job yesterday, and the boss told me to stay after work to fix it but was real cool about it. Glad I was told"

   c. "Boss said I did some little thing wrong and I've got to put in overtime to fix it, and I don't see why"

   d. "Boss said I did something wrong, and I pointed out that if I couldn't do it right the first time, I probably won't get it right the second time. Pretty savvy of me to think that one up, huh?"

3. You are quietly working away when two new co-workers begin to argue nearby. Then they start yelling louder, then screaming, and then it looks like they're about to come to blows. Your best response is to _____.

   a. go place yourself in between them and say, "You can't fight here unless there's a referee … and I'll do it if you want me to"

   b. ignore them; if they want to beat each other, it's their business

   c. go get some of the other workers and watch the fight

   d. go over to the two new workers and calmly say, "I'm sorry for butting in here, I know it's none of my business, but it looked like you two were about to start fighting; you know you'll get fired for that, don't you"

4. You and J.J. are working on a two-person job that requires you both take lunch at the same time. J.J. needs to get to the bank by noon, so needs to take lunch from 11 A.M. to noon. You need to take a mid-afternoon lunch: your car is in the shop, and you've arranged with a friend to pick you up around 2 P.M. to take you to the mechanic's, which closes at 4 P.M. (The mechanic opens up again tomorrow morning, Saturday, for a few hours.) The best compromise would be _____.

   a. you both take lunch from 11 to noon, and then you and J.J. ask the boss if you both can leave at 3:30, so that J.J. can drive you to pick up your car before 4

   b. J.J. goes to the bank now and you cover until J.J. gets back, then you go pick up your car this afternoon and J.J. covers for you; no one will probably notice, and you'll both work like dogs to build in some slack time when one or the other of you is absent

   c. you both take lunch at 11, and you lend J.J. some money for the weekend and J.J. agrees to pay you back on Monday

   d. you agree to take lunch later: J.J. goes to the bank, you call your friend to see if you can get that trip to the mechanic on Saturday morning, and then J.J. drives you home after work

5. You and a co-worker, R.N., aren't getting along. You both go to see the boss, who gives you five minutes to come to an agreement. You say it started when R.N. stole one of your hammers. R.N. says it started when you made fun of how R.N looked. You reply, "Your appearance is a safety hazard, wouldn't you say, boss?" R.N. responds, "What about you? Coming in late every day this week, have you noticed that, boss?" The boss says that neither one of you is perfect and to go work things out among yourselves. What should you do next?

   a. Steer clear of R.N., who's liable to say anything behind your back. You also let your co-workers know that R.N. is a proven tattletale.

   b. Confront R.N. about bad-mouthing you in front of the boss, say that R.N. crossed the line once and for all, and that the time has come to apologize.

   c. Let things cool off, and then try to talk to R.N. the next day by saying, "You know, we can't let this go on, it's gonna hurt us both, not to mention the whole crew. Let's think of them for a change, instead of each other."

   d. Tell R.N. it's water under the bridge—even though you know it isn't—and that maybe you two should go out for a beer to try to patch things up.

6. It seems your boss always makes sure to point out your mistakes in front of the rest of the crew. This morning at the tool box talk, while showing the crew how to do something, the boss says, "And don't do it this way, like numbskull over there did the other day," while pointing at you. You feel hurt and embarrassed. Later, you ask the boss about it, who replies, "If you can't take the heat, get out of the kitchen." What is your best course of action?

   a. Tell the boss then and there to take this job and shove it, and then quit; you'll get another job and no future employer will hold it against you that you quit, given the circumstances.

   b. Ask yourself whether you might not be better off looking for another job; if you really like this one and don't want to leave just because the boss isn't treating you well, either let the matter go or consult the employee handbook on whether the boss's treatment of you could be grounds for filing a grievance.

   c. Ask a co-worker if this is the kind of behavior you can file a grievance about.

   d. Wait until you can catch the boss doing something wrong, even if it takes months, and then tell your boss's supervisor as soon as it happens.

7. Your co-worker is annoying, always complaining about how hard the job is and about family members who are a pain in the neck. You've never paid much attention, hoping the co-worker will get the message and stop. But the co-worker seems to be actively seeking you out these days as a sounding board. You finally decide it's time to say something. What should it be?

   a. "I'm not here to listen to your problems, dig? I'm not your shrink!"

   b. "Speaking of your family, if I were a part of it, I'd have left a long time ago to get away from your complaining … I'm only joking, but please, shut up."

   c. "Look, I don't mind hearing some griping every now and then—I know I do it—but I just can't concentrate on my job when you keep telling me this stuff all the time. Wouldn't it do more good to talk to your family about family issues or to the boss about work stuff, instead of talking to me?"

   d. "You know, I'm really, really, really, really, really, really, really, really sorry. I mean, really sorry. Did I say it enough? Now, let me get back to work!"

8. You are a new worker, doing your best: always offering to help, trying to be a good teammate. A more experienced worker is a bit of a bully; this person is not your boss but acts that way toward you a lot. One day the co-worker tells you to do part of that co-worker's job. You ask why. The co-worker responds, "Don't argue, kid; the boss put me in charge of you, so you do as I say. You don't, I say one word to the boss, and you're out on the street." What you say in response?

   a. "Get lost."

   b. "I certainly will … boss. And tomorrow, just to make sure of the chain of command, I'll go tell my old boss that I understand you're my boss now because you say so. I'm sure the boss will be glad to know how you're delegating authority around here."

   c. "Yes, boss. You want me to wash your car for you, after I'm done?"

   d. "You're not my boss, OK? You want to give me advice or tell me I'm doing something wrong, fine; I know I've got a lot to learn. But I don't take orders from you unless I hear differently from my boss personally. If you want me to take orders from you, go tell the boss to tell me in person that I now have to take orders from you. Until I hear otherwise from my boss, I'm not taking any orders from you."

9. Your boss tells L.T. to cut and bend sheet metal for ductwork. L.T. does not like doing this task and asks you very nicely if you could do it instead. What is your best course of action?

   a. Do as L.T. asks because the request was made politely.

   b. Lie and say that you don't know how to do it.

   c. Say that you have your own work to do and your own responsibilities, and that the boss told L.T. to do it, not you.

   d. Agree to do the task, but then don't do it, so L.T. will get in trouble with the boss and learn a lesson.

10. T.J. is one of the more experienced workers on the job and knows everything about everyone. T.J. passes along gossip whether the information is true or not. The best way to deal with T.J. is to _____.

    a. be careful what you say to T.J., and change the subject when T.J. wants to gossip

    b. explain to T.J. why gossiping in the workplace is not a good idea, and give a few examples of past gossip that has caused problems

    c. set a trap by giving T.J. phony information about yourself and others

    d. tell your boss about the damage T.J. is causing by spreading gossip

Individual Activities

Activity 1: Using Words That Won't Put Others on the Defensive

In this module, you've learned that using the word *you* in conflict situations could make the other person defensive. Instead of saying *you*, it is better to say *I*, *we*, *the company rules, manufacturer's guidelines, OSHA,* or *the code*. Talking this way may feel a little awkward at first, but it'll pay off down the line when you're discussing possibly delicate matters with co-workers. This activity will give you a chance to practice. Read the following examples and change them so that your listener is not put on the defensive.

1. *Poor approach:* "It takes you two days longer than everyone else to do the same job."

   Better approach: _____

2. *Poor approach:* "You're always criticizing me and finding fault with what I do."

   Better approach: _____

3. *Poor approach:* "Are you crazy? Can't you see that cord's frayed? Do you rookies have to be told how to do everything?"

   Better approach: _____

4. *Poor approach:* "You installed that water heater all wrong. Can't you see you forgot the bracing?"

   Better approach: _____

5. *Poor approach:* "You are really sloppy on the job. Your work area is a pigsty."

   Better approach: _____

Activity 2: Seeing the Positive Side of Conflict

Conflict is about as common as agreement in everyday life. We cannot always avoid conflictsometimes conflict can be beneficial. Think about some conflicts you've had with others recently. Did anything good happen as a result of those conflicts? Did you learn anything? Was a problem solved? Did it give you a chance to understand someone else's point of view? Write down positive outcomes of conflicts you have experienced.

1. _____

   _____

2. _____

   _____

3. _____

   _____

Five-Step Conflict-Resolution Process		
Step 1	Bring the conflict into the open.	
Step 2	Discuss and analyze the reasons for the conflict.	
Step 3	Develop possible solutions. Remember that all workers involved should collaborate and compromise.	
Step 4	Choose and carry out a solution.	
Step 5	Evaluate the solution.	

Activity 3: Resolving Conflict

Your goal in this activity is to resolve a conflict using the five-step conflict-resolution process you learned in this module. After you've read and thought about the situation, work with two or three of your classmates to complete the conflict-resolution grid. However, before you start completing the grid, set up some ground rules. Recall the ground rules you read about in this module, such as treating one another with respect and letting others speak without interruption.

Team Members

1. _____   2. _____

3. _____   4. _____

Situation: Two construction crews have been assigned to a new project. You are on one of the crews. To keep the job moving, the crews take staggered lunch breaks. Each day, both the drink and snack machines are completely empty by the time the second shift is ready to take its break. The workers in the second shift are upset about this situation, and now conflict and bad feelings have sprung up between the shifts.

Ground Rules

1. _____

2. _____

3. _____

4. _____

5. _____

To complete this activity, follow the conflict-resolution steps.

Discussion questions

1. How can you bring this conflict into the open?

2. Is collaboration possible?

3. Is compromise possible?

4. What is the common ground? That is, what outcome will benefit all of you?

5. Once you've resolved the problem, how will you go about making sure that the conflict doesn't arise again?

Activity 4: Defining Success in Conflict Resolution

Most of us define success as winning. That definition is fine for competitive sports, where there must be a clear winner and a clear loser. When you resolve conflicts at work, however, you must define success differently. Success at work means that all parties feel satisfied so that work can continue without a lot of drama and tension.

In this activity, you and two or three of your classmates will review some typical job conflicts and then state whether you think the outcome is successful. Remember that in the workplace, conflicts are best resolved with collaboration and compromise. When everyone wins a little, the entire team wins a lot.

Team Members

1. _____   2. _____

3. _____   4. _____

Discussion questions

Read and think about each situation. Then answer the following questions:

1. Was this conflict resolution successful? Why or why not?

2. What are the possible good outcomes of this resolution? What are the possible bad outcomes?

3. If the resolution was not successful, what could have been done to change the outcome?

Situation 1: Your co-worker is messier than you are. You can't stand seeing stuff all over the place. You tend to nag your co-worker about the messiness. You talk things over, and your co-worker agrees to pick up the area at least twice a week. You agree to learn to live with a little more messiness than you are used to and to stop nagging.

Successful resolution?   ◯ Yes   ◯ No

Situation 2: You and three co-workers agree to start a carpool to save money. You have the biggest vehicle and volunteer to drive. In exchange, your co-workers agree to pay for gas and parking fees. After awhile, you get tired of being the chauffeur, and complain. Your co-workers offer to alternate driving your vehicle or to drive their own vehicles, but you don't want to do either of those things. Your co-workers' cars are too small, and you don't want anyone touching your car. Your attitude makes everyone mad at you, and you get mad at everyone else. The carpool breaks up.

Successful resolution?   ◯ Yes   ◯ No

Situation 3: You are a man and are put on a team that includes two women. You make several disrespectful remarks about women in construction. You think you are just being funny, but the women don't appreciate your humor and are very cold toward you. You decide to apologize for your remarks and to be politically correct when the women are around. Once they are out of earshot, however, you revert to your old ways.

Successful resolution?   ◯ Yes   ◯ No

Situation 4: You work on the second shift. When you and your crew come in, the snack area is always a mess, with dirty cups, spilled condiments, and empty snack containers scattered about. Often, one of the coffee pots has a burned-on film, and supplies have been used up and not replaced. You and your crew talk to the crew on the first shift and ask them to leave a clean snack area and to replace supplies. That's when you learn that they have to deal with your crew's mess at the start of their shift. You all agree to come up with a plan that will ensure that someone on each crew cleans up the area at the end of the shift and replaces supplies.

Successful resolution?   ◯ Yes   ◯ No

Situation 5: Your co-worker is a workaholic who comes to work early and stays late. You believe that coming in to work on time and leaving on time are fine. You put in a good day's work each day, but your co-worker is constantly on you because you don't share the same work ethic. Fed up with this, you confront your co-worker one morning and say, "It's great that you love work so much. But I am sick of your bragging about what a great worker you are, and so is everybody else. Back off or you will get what's coming to you!" Your co-worker leaves you alone after that.

Successful resolution?   ◯ Yes   ◯ No

Activity 5: Your Turn—Talking to the Boss

In this activity, you and two or three of your classmates will act as the scriptwriters. Study each boss–worker conflict situation. Based on what have you learned in this module, write down what you think you should say.

Team Members

1. _____   2. _____

3. _____   4. _____

As you write your remarks, consider these questions:

1. Can I just walk up to the boss with this situation, or should I ask the boss if this is a good time to talk?

2. What can I say to keep from sounding negative or defensive?

3. What can I say to keep from sounding threatening?

4. How can I show the boss that I want to contribute to the solution?

5. Should I be asking the boss about this at all?

Situation	Sample Remarks
1. You've been late twice this week. The boss says you have to work overtime to make up for the lost time, but you play on a softball team and don't want to miss two scheduled games.	
2. You are upset because you feel that the boss has passed you over for promotion.	
3. You are upset because the boss criticized your work and made you redo it.	
4. You feel like you are doing more work than everyone else.	
5. You want the company to pay for some computer courses you want to take.	
6. You want the boss to change your work schedule so that you can take some classes.	
7. You want the boss to help get a bullying co-worker off your back.	
8. At a crew meeting, the boss makes a blanket statement about "some workers" being late. You feel that the boss took the wrong approach. You want to tell the boss that the workers who come in on time are upset.	
9. The boss made a hasty decision that you know will cause problems. You must tell the boss about these problems.	
10. Although your company lost a job to a competitor recently, you need to ask for a raise.	

Activity 6: You, The Boss, or Forget About It

In this module, you have learned there are certain conflicts you should handle yourself, certain situations when the boss can or should be involved, and certain disputes that you should just walk away from or forget about. On the jobsite, these choices will not always be obvious, and the same situation may need to be handled differently at different times depending on the circumstance. This activity gives you a chance to think about a conflict before you charge in to fix it or run to the boss with the problem. Working with two or three of your classmates, complete the following grid. Read each situation and check one of the following:

— *You,* if you should handle the conflict on your own.

— *The Boss,* if the supervisor should be involved with or handle the conflict.

— *Forget About It,* if it is a situation you should walk away from or ignore.

Team Members

1. _____   2. _____

3. _____   4. _____

	Situation	You	The Boss	Forget About It
1.	A co-worker keeps borrowing money from you and forgets to pay you back.	○	○	○
2.	A co-worker threatens to beat you up after work.	○	○	○
3.	You see a co-worker stealing tools from the worksite.	○	○	○
4.	You have to take orders from someone younger than you are.	○	○	○
5.	Your co-worker is sexually harassing you.	○	○	○
6.	Your co-worker tells obscene jokes you find offensive.	○	○	○
7.	Two of your co-workers are constantly bickering.	○	○	○
8.	You are in a carpool and because one of its members is habitually slow in the morning, you run late at least twice a week.	○	○	○
9.	Your co-worker is openly rude and hostile to you.	○	○	○

	Situation	You	The Boss	Forget About It
10.	Your co-worker often collects money for various charitable causes. You don't want to contribute and think that money should not be collected on company time.	○	○	○
11.	You are having problems with your family, and you feel annoyed and irritated at work.	○	○	○
12.	Your co-worker is a whiz at technology and makes you feel like an idiot.	○	○	○
13.	Your co-worker hates the job and complains about it every day.	○	○	○
14.	Your co-worker says you have body odor (you do).	○	○	○
15.	Your co-worker says you have body odor (you don't).	○	○	○
16.	Your co-worker likes to hum softly while working, which makes it hard for you to concentrate.	○	○	○
17.	Your co-worker got a raise and you did not.	○	○	○
18.	Your co-worker is a mean-spirited gossip.	○	○	○
19.	Your co-worker tries to control everything and everybody, including you.	○	○	○
20.	Your co-worker is from another country and struggles with English.	○	○	○

Activity 7: True or False? Test Your Perceptions

A lot of conflict is caused by the perceptions or beliefs we have about a situation or the people involved. In this activity, you will first read the statements and decide whether they are true or false on your own. Then compare your answers with your team members' answers.

Team Members

1. _____   2. _____

3. _____   4. _____

Discussion questions

1. Did all of you choose the same answers?

2. Talk about the answers you differed on.

3. Did this exercise change your mind about anything?

	Statement	True	False
1.	People who are older than I am know more about every construction topic than I do.	○	○
2.	I know more about technology and computers than people who are older than I am.	○	○
3.	People who dress differently than everyone else are just strange.	○	○
4.	Women should not have tattoos.	○	○
5.	Men should not get manicures.	○	○
6.	Workaholics think they are superior to the rest of us because they work so much.	○	○
7.	A boss just doesn't understand what it's like to be a worker.	○	○
8.	Construction workers usually discuss and handle their problems with their fists.	○	○
9.	Older workers are not supposed to get along with younger workers.	○	○
10.	People from different backgrounds can't work together in construction, no matter what.	○	○

Giving and Receiving Criticism

"It's nothing personal, Sonny… it's strictly business."

– Michael Corleone,
The Godfather

Introduction

As you progress in your construction career, you will likely make mistakes, which means you could be criticized. Being a part of the construction industry means learning new things all the time and keeping up with new developments. When you are learning new things, you are bound to make mistakes and receive your fair share of criticism. You need to learn how to handle criticism, and the first thing to know is that criticism should not be taken personally.

Criticism can be delivered in different ways. It can be mild or harsh. Don't get upset about the criticism, no matter how it is given. Think about what is being said, and decide if the criticism is valid. If it is, what you are going to do about it? Think about what you can do to take positive action to avoid similar problems in the future. The important thing to remember is that you can learn from criticism in any form.

Sometimes you will be the person doing the criticizing. For example, you may be assigned to train a new worker who makes typical beginner's mistakes. You need to know how to point out those mistakes in a positive way, without being harsh or insulting, so that the new worker can focus on improving. You want to help, not condemn.

The toughest criticism of all, perhaps, is the criticism you direct at yourself. You might be much harder on yourself than others are. It is important that you know how to recognize and admit you have made a mistake and to hold yourself accountable for it, but being too hard on yourself can prevent you from reaching your full potential. In this module, you will learn how to receive criticism from others and how to give criticism, both to others and to yourself.

What is criticism?

We often associate the word *criticism* with something negative or unpleasant; criticism, however, should not be considered a bad thing. Delivered properly, criticism is a valuable and necessary learning tool. In fact, the origin of the word is related to analyzing; like literary criticism, for example. The goal of true criticism is to help someone do his or her best; criticism should not make someone feel bad or guilty. The intention of criticism is not humiliation. Proper criticism means those who have received it understand what they have done wrong and are inspired to improve.

There are two types of criticism:

- Destructive criticism is hurtful and insulting. It strips people of their self-esteem, makes them feel worthless, and creates bad feelings. It has no place on the job or in life. "You're a crummy worker" or "You're a moron" are examples of destructive criticism.

- Constructive criticism is tactful and supportive. It helps others to improve. It is the only type of criticism that is useful, and it should always be used on the job and elsewhere. "You're trying to force that tool because you're in a hurry; relax, and let me show you how to do it more effectively and so you won't break the unit," is an example of constructive criticism.

Benefiting from constructive criticism

Most supervisors give constructive criticism regularly, because their goal is to help their workers improve. Correcting your mistakes can mean the difference between success and failure or even life and death; therefore, you need to be able to take criticism well so that you can benefit from it and improve your skills. To get the most out of constructive criticism, keep these points in mind:

- **Anticipate criticism and be aware of the source.** It is easier to deal with criticism if you expect it and if you think of it as an opportunity to learn and improve your skills. Keep in mind that your immediate supervisor is responsible for your work.

- **Don't take criticism of your job performance personally.** Recognize that your supervisor is criticizing your mistakes and not you personally. Get in the habit of thinking, "This isn't about me, it's about the job I'm doing, and I can learn how to do that better." Instead of being bothered by your boss's criticism, ask for specific details on what you need to correct.

- **Look at your work from your supervisor's point of view.** Your boss is obligated to criticize workers when their work or behavior on the job is shoddy or poor. Part of knowing how to benefit from constructive criticism is being able to honestly evaluate your work. If your boss gives you an earful about how poorly you did a task or how you need to be a better teammate, you should admit the criticism is justified and then start concentrating on improving your performance.

- **Don't get defensive.** When your boss criticizes you, your first reaction may be to defend yourself, respond sarcastically, or act sullen. Resist those temptations and keep an open mind. Supervisors will think more highly of you if you demonstrate that you can take criticism well. Even if your boss is really chewing you out, hold your temper.

- **Don't make excuses.** You may want to explain your mistake or shift the blame to somebody else. Don't. It is much better to take responsibility for your mistakes and move on.

- **Ask for clarification and specifics.** Some criticism is vague. Don't be afraid to ask questions if you do not fully understand what you did incorrectly. Ask those questions respectfully. Remember that your goal is to improve your performance, not defend your mistakes.

- **State your plan for improvement and carry it out.** There is no better way to impress your boss than to take criticism seriously. Tell your boss what you will do to correct your mistakes. For example, if your boss reprimands you for being late, you could say, "I'm getting a better alarm clock so that I won't oversleep again." You must respond to criticism with not only words but action, so be sure to get that new alarm clock.

- **Don't allow criticism to slow you down.** Remember, it is how you handle criticism that matters. Don't allow criticism to lower your self-esteem. Think of constructive criticism as the boss thinking highly enough of you to analyze your work in the first place.

- **Understand your right to disagree when the criticism is not warranted.** After carefully evaluating the criticism you receive, you may feel it is not justified. Sometimes you are right and the other person is wrong. If you think that is case, offer your side of the story respectfully and clearly, and give examples to support your point of view.

Dealing with destructive criticism

People give destructive criticism for a number of reasons. Sometimes people are rude or inconsiderate of others' feelings. Sometimes people only know how to give destructive criticism because that's all they have ever heard in their lives. People may blurt out hurtful and disrespectful things as they are giving criticism because they let their emotions get the better of them, but then they apologize later. Some people might be aware they are giving destructive criticism but prefer to do it that way, simply because they are not very nice people. When people are leveling destructive criticism at you, there are ways you can handle it so that you don't get upset.

- Control your emotions while you are listening. If you respond angrily to the person giving you destructive criticism, that will probably only stir up that person even more, making the situation even more tense or hostile.

- Step back from the situation. If the person pauses, you don't have to answer right away, or you can say something like, "Go on, I'm listening." If you respond calmly and without aggression to those who are hostile and insulting, it might catch them off guard. They might just be trying to rattle or antagonize you. If you keep your cool, they might respond by making more constructive and less abusive criticisms.

- Remember that you can respond constructively to destructive criticism. If there is merit to the criticism amid all the insults, focus on improving and not on how the criticism was delivered.

- Ask the person to give you constructive criticism instead, and tell the person how destructive criticism makes you feel.

- If the criticism is not true or valid, do not dwell on it. Move on.

Example: Your boss has had a frustrating day and yells, "Why can't you do anything right? I have to watch you every minute. You are so dumb!"

Your response: Take a deep breath and calm down. Think. Of course you can do a lot of things right. The boss doesn't really watch you every minute, and you know you are not dumb. You could say, "Instead of yelling at me, take a minute and show me how to do it right."

Giving constructive criticism

Giving constructive criticism is a skill that takes time and patience to develop. To offer criticism that is beneficial and helpful, and to do it without being insulting or hurtful, follow these guidelines:

- **Decide what is worth criticizing.** First off, determine if you should be criticizing your co-worker at all. For example, if a co-worker isn't wearing a hard hat, you should definitely speak up; that is a safety risk. However, if you don't like how a co-worker is installing cabinets, hold your tongue. Work techniques can differ without affecting quality. Avoid criticizing personal preferences, such as in music or politics; these have nothing to do with getting the job done safely and well. Don't criticize someone's appearance unless it affects safety, but if someone's hair, jewelry, or clothing might be detrimental to safety or performance, then you should say so—tactfully.

- **Choose the time and place wisely.** Never criticize people in front of their peers. It will embarrass them and make them resent you. Take them off to the side, and choose a time when they are feeling less stressed or pressured. If the person is having a bad day or is under a lot of stress, and if the criticism can wait, put it off for a day or two.

- **Choose your words carefully.** Don't swear or call people names. Don't sound angry, frustrated, or sarcastic. Stick to the facts and stay clear of personal comments.

- **Criticize the behavior, not the person.** To prevent hard feelings, make it clear that you are criticizing the mistake, not the person. Be specific about what the person has done (or not done). Consider the following example:

Example: A co-worker leaves the jobsite each day without putting tools away.

Destructive criticism: "How many times do I have to tell you to put the tools away before you leave? Can't you understand simple directions? What are you, deaf, dumb, or both?"

Constructive criticism: "It's important to put the tools away at the end of the day. You do need to start doing a better job of that, you know. We insist on that around here, and when tools are not put away, it causes problems for everyone. It only takes a few minutes, and it's a big help to everyone."

- **Decide how many things to criticize at one time.** Usually, it is best to focus on one issue at a time. This allows you to be detailed in your criticism. Sometimes, however, you may have to bring up a number of issues at once. If this is the case, let the person know up front. Focus on the present, and don't bring up old issues. It is important to point out improper behavior or incorrect techniques the first time they occur; so, even if you have other criticisms, do not neglect to discuss these. Reminding your co-worker of your company's standards, and how the whole team is counting on them to abide by the rules, is an excellent way to give constructive criticism.

    Example: You are teaching a nervous trainee how to install kitchen cabinets.

    Destructive criticism: "You are messing up a lot! Look at all the mistakes you made just today! First, you racked that cabinet. Then you installed those bases out of plumb, and I can't believe your finished work. Look at all those nail holes!"

    Constructive criticism: "Slow down. You're trying to get too much done too fast. You've made three mistakes. Let me show you how to do these jobs the right way."

- **Offer suggestions.** Remember the importance of offering solutions and not just complaints. With constructive criticism, that means making suggestions on how to do something better, and not just focusing on what someone did incorrectly.

    Example: You and a co-worker are cutting wood to various lengths. You notice that your co-worker, who measures one time and then quickly cuts the wood, is cutting the wood to the wrong lengths and mixing up installation pieces with waste pieces.

    Destructive criticism: "What's the matter with you? Can't you even do a simple thing like cutting wood correctly? Look at all this waste!"

    Constructive criticism: "Here's a tip on doing that. To make sure you've measured correctly, measure twice before you cut. Then put an X on the waste piece."

- **When possible, offer praise along with the criticism.** Commend more, and don't condemn so much. People are much more receptive to criticism if it comes with a nice word or some praise. It may not be possible to praise the person's work, but try to find something positive to say, even if it's about the person's effort or enthusiasm.

    Example: Your co-worker is struggling to make angle cuts in molding, has made several mistakes, and keeps trying to get the angles right.

    Destructive criticism: "Look at this mess! You are never going to learn to do this right!"

    Constructive criticism: "I admire your determination, but I think you are too eager to start cutting. Take the time to measure accurately and think through the steps before you turn on the saw. Let me show you how."

- **Follow up.** To make sure that problems are not repeated, check up on the situation later on; it provides a good opportunity to offer your ongoing support.

- **Learn how to criticize your own work evenly.** Evaluate your work based on your experience and your company's standards. Don't tell yourself that you are better than anyone else. But don't be too critical of yourself, either. A good balance is to take pride in your work while acknowledging you have a lot to learn and that every day is an opportunity to improve your knowledge and add to your skills.

Summary

Criticism can be destructive or constructive. Destructive criticism is hurtful and leads to bad feelings and tension. Constructive criticism is all about helping people to improve by pointing out their mistakes calmly and respectfully, and offering suggestions on how to improve performance and behavior. Constructive criticism is far more useful and helpful than destructive criticism, but you will experience both in your career. You will need to know how to get the most out of constructive criticism while not allowing destructive criticism to get you down or make you mad. When you have to criticize someone, always do it constructively by focusing on the mistake and not the person who made it, offering suggestions, and giving it at the right time and place.

Here's a quick quiz that allows you to apply what you've learned in this module. Select the best possible answer.

1. You hear the boss say to one of your co-workers, in front of everyone, "That was so stupid! I can't believe you mixed that cement before we were ready for it. It's all going to go to waste! You just don't pay attention to your work." If you were asked your opinion, what would you say?

   a. "The boss is right and handled the criticism appropriately."

   b. "Everyone makes mistakes, and the boss should have let this one go."

   c. "This is constructive criticism. The boss has to make sure everyone understands how wasteful the mistake was, which is why it was said in front of everyone and so vehemently."

   d. "This is destructive criticism. The boss could have handled the situation in a more positive and subtle way."

2. You are fastening deck boards to joists and you strip the threads on several screws. Your boss says, "Listen, this is not a huge deal. I'll show you how to fix that mistake, and then show you how you should change your technique so that you don't keep stripping threads." If you were asked for your opinion of this situation, what would you say?

   a. "This is constructive criticism. The boss handled the situation well."

   b. "This is constructive criticism—the type a good worker like me deserves. With some of my co-workers who aren't as good are I am, however, the boss would have to be sterner."

   c. "This is destructive criticism. The boss made me feel stupid by saying that this was not a huge deal."

   d. "Whether destructive or constructive, I deserved the criticism in the first place for making such a dumb mistake. I can't even use a screw without stripping it!"

3. The site supervisor tells you to move some scaffolding out of the way, and you do it. A few minutes later, your immediate supervisor criticizes you somewhat harshly for moving the scaffolding. What is your best course of action?

   a. Keep quiet and let your boss blow off some steam; you know you have nothing to apologize for.

   b. Say that your boss is out of line because you were just following the site supervisor's orders.

   c. Apologize for not checking with your boss before you did what the site supervisor said, and ask what you should do in the future if the situation comes up again.

   d. Apologize to your boss, then track down the site supervisor to explain what happened.

4. Your boss takes you aside and points out several errors you made in framing a room. The boss shows you how to fix the errors and mentions that you should concentrate more on your work and spend less time joking with co-workers. What is your best response to this situation?

   a. Decide that your boss is critiquing your work and work habits, not you personally.

   b. Decide that your boss likes you as a person despite the comments about your work and joking.

   c. Feel that your boss has it in for you and should cut you some slack.

   d. Feel that the comments about the framing mistakes are fair but that the comments about joking with the co-workers aren't.

5. It's the end of the day, and you're getting ready to leave. Suddenly, one of your co-workers comes up to you and screams, "Yo, where you off to? You'd better stay and fix up that mess you left behind because I'm not staying to cover for you, understand?" You don't have any idea what your co-worker is talking about, so you _____.

   a. ignore this person, punch out, and leave

   b. point out that your co-worker is not your boss and cannot tell you what to do

   c. yell back demanding proof that it was you who messed up

   d. ask them to stop yelling and explain the problem

6. A more experienced co-worker tells you that you're hopeless and too stupid to ever become any kind of worthwhile construction worker. What is your best course of action?

   a. Say, "Yeah? So if you're so smart, why aren't you the boss instead of doing what I'm doing!"

   b. Stay calm. Explain that you know you've got a lot to learn but that doesn't make you stupid, and that calling you stupid is not going to help you to learn.

   c. Stay calm. Realize this co-worker has a bad reputation for having a foul temper, and is probably just saying this because you are showing them up with your work ethic and energy.

   d. In a calm and respectful voice say, "Don't ever call me stupid again."

7. You're training a new worker, B.K., who seems smart and efficient but who is heavily tattooed. You can't understand why people get tattoos; the whole idea kind of creeps you out. What is the proper reaction to the fact B.K. has all those tattoos?

   a. There's no reason for any reaction. The tattoos have nothing to do with the job or B.K.'s ability to do it.

   b. Joke, "B.K., all those tattoos make you look like a thug."

   c. Say, "B.K., I'm cool with the tattoos, but the boss has a problem with people who are covered with them, so I'd advise you to get them removed if you want to keep your job."

   d. Say, "Those are the ugliest tattoos I've ever seen. If you're going to cover yourself up with them, at least get someone to do a good job of it. Right now, you look like something out of a junkyard."

8. You're training a few new workers, and one of them, C.C., keeps making mistakes. You must point out these mistakes. What is your best course of action?

   a. Talk about C.C.'s mistakes publicly, not to single out C.C. personally but just to save time overall by discussing the mistakes as C.C. makes them.

   b. Talk to C.C. about the mistakes in private, and offer some advice for avoiding them in the future.

   c. Say, "C.C., I can see you're trying as hard as you can, but you're making an awful lot of mistakes, which tells me this technique is too advanced for you."

   d. Tell C.C. in front of the others to get with the program and stop making mistakes that any six-year-old could avoid.

9. You notice that J.J., a trainee, has made five mistakes: one major and four minor ones. The minor mistakes aren't related to the major mistake. You decide to talk to J.J. about the mistakes. What is your best strategy?

   a. Talk to J.J. about all the mistakes, because bad habits need to be nipped in the bud.

   b. Discuss the major mistake first, and once that's resolved, bring up the other ones.

   c. Mention the minor mistakes at the worksite and talk about the major mistake later in private.

   d. Take a wait-and-see approach. J.J. may improve, and you can avoid having to give criticism.

10. P.J., a more experienced co-worker, has been teaching you a difficult new glazing technique. P.J. has been somewhat abrupt when pointing out your mistakes and hasn't given you any encouragement, but P.J. hasn't been insulting or disrespectful, either. Just all business. Now you're ready to try the technique on your own. What should you tell yourself, as you get ready to begin the job?

   a. "There's no way I'll ever be any good at this, which is why P.J. was so abrupt with me during the training."

   b. "I've got this stuff down cold, better than P.J. probably. When I teach it to someone, I'll be nicer about it, too."

   c. "I know how to do this task now. I'll work carefully and ask P.J. for advice if I run into problems."

   d. "I won't worry too much about making mistakes. If I do, I'll just say P.J. was a lousy teacher and never had a nice word to say to me. After all, that's only the truth."

Individual Activities

Activity 1: Responding to Criticism

To complete this activity, read each criticism and decide whether it is constructive or destructive. Then, based on what you learned in this module, respond to each criticism. Be sure to write what you should say and not what you might want to say in the heat of the moment.

	Criticism	Source of Criticism	Constructive or Destructive?	Your Response
1.	"You're not a team player. You come in late and leave a mess behind at the end of the day at your station."	Boss		
2.	"After that mistake I made, I know everyone hates me. I'll never be part of this team."	Boss		
3.	"You're really slowing me down. You're too young, and you don't know what you're doing."	Co-worker		
4.	"Hold on! It's not safe to use that chisel. See? The end is all flattened out."	Co-worker		
5.	"You need to pay closer attention when you set up the saw for angle cuts. I'll show you a good way to do that."	Boss		
6.	"I know I can be a better carpenter. I'm going to ask the boss about that training course."	You		
7.	"Those goggles you've got are piece of junk. Why don't you get some new ones? Too stupid to see how important it is, literally?"	Co-worker		

Continued

Criticism	Source of Criticism	Constructive or Destructive?	Your Response
8. "Why can't you figure out what you really need? I'm sick of taking stuff back. Get your act together!"	Supplier		
9. "This is the third day in a row that you've been late. That means you're lazy and unreliable."	Boss		
10. "I've been meaning to give you these breath mints. Take one, or maybe 12. Every time you come back from lunch, your breath stinks up the joint something fierce."	Co-worker		

Activity 2: Converting Destructive Criticism into Constructive Criticism

To keep from giving destructive criticism to others, you must think before you speak. To complete this activity, read the following situations and the accompanying destructive criticism. Write out how you would turn the destructive criticism into constructive criticism. The first one has been completed as an example.

	Situation	Destructive Criticism	Constructive Criticism
1.	You're teaching a trainee to mud and sand drywall joints. Your trainee is too heavy-handed with the sander.	"Yikes! Can't you watch what you're doing? Just look at this mess. Just get out of the way—I'll fix it."	"Ease up. You don't have to wear yourself out. Let the tool do the work. Like this, see? Give it another try and go easy."
2.	One member of your team calls in sick a lot, takes long lunches, and often leaves early.	"You're a lousy worker, and we're all getting sick of you. We'd be better off without your useless presence."	
3.	A co-worker cuts a two-by-four at the wrong angle. You are out of two-by-fours, and now you can't finish the job today.	"You know, you really messed me up. Now I gotta do it all over again because you can't use a simple saw. And you're in construction?"	
4.	You can't seem to get the hang of a power drill. You've stripped some screws, and you've buried a drill bit in a beam.	"I'm so stupid. Why can't I get this? I bet everybody thinks I'm an idiot. None of them seems to be having any problems."	
5.	Your co-worker keeps misplacing or losing tools and supplies you need to get a job done.	"I think you'd lose your head if it weren't attached to your body. I just wish you'd lose your way to work someday and never come back."	

Activity 3: Your Action Plan for Improvement

As you have learned, one good way to deal with constructive criticism is to state and carry out an action plan for improvement. Most experts agree that writing down a plan is an important first step. To complete this activity, recall something your boss has criticized you about recently. Think about the following questions:

1. What am I doing incorrectly?

2. How much time do I think I will need to improve?

3. Do I need any tools, equipment, or training to help me improve?

Once you have thought about these questions, fill in the Self-Improvement Grid. In the first column, write down a constructive criticism your boss has given you recently. In the second column, write out your action plan, how long it will take you to carry it out, and whether you will need tools, equipment, or training to accomplish that goal.

Self-Improvement Grid

Problem Statement	Action Plan
My boss has told me that I	To correct this behavior, I will
	To succeed, I might need

Activity 4: Giving Fair Criticism

Read the following case studies. Then in groups of three or four, discuss the questions that follow.

Team Members

1. _____ 2. _____

3. _____ 4. _____

Case Study 1

T.R., who works on your team, is a neat and reliable plumber and also does the following things:

- Is often about 10 minutes late at the start of the workweek
- Hums softly while working
- Takes a little longer than the other plumbers to dry fit pipes
- Likes to read at lunch and is sometimes late getting back to work
- Has a habit of hoarding supplies
- Has a habit of swearing at the pipes and fittings
- Uses nicknames that some of the team find offensive
- Wears a bandanna with drawings of parrots on it

Discussion Questions

1. Which of T.R.'s behaviors must the boss handle? Which of T.R.'s behaviors should the team handle?

2. Which of T.R.'s behaviors should the team not bring up?

3. Choose one of the behaviors you think the team should handle and discuss how you would constructively criticize T.R.'s behavior.

Case Study 2

M.L. is nursing a broken heart and is having money troubles. Normally reliable, M.L. can't seem to concentrate on anything and keeps making mistakes that set everyone back. This absentmindedness is becoming a safety issue. In the last week, the crew has noted the following in M.L.'s behavior:

- Is drinking more coffee than usual

- Has started to wear a T-shirt with "Life Sucks" printed on it

- Keeps forgetting to wear safety gear

- Keeps asking for instructions to be repeated

- Often sits and gazes off into the distance

- Is not paying attention when acting as a safety watch

- Has been giving the wrong hand signals to the crane operator

Discussion Questions

1. Which of M.L.'s behaviors must the boss handle? Which of M.L.'s behaviors should the team handle?

2. Which of M.L.'s behaviors should the team not bring up?

3. Choose one of the behaviors you think the team should address, and discuss how you would constructively criticize M.L.'s behavior.

Activity 5: Your Turn

In this module, you have gained a lot of practice in dealing with destructive criticism and in giving constructive criticism at work. Working with a group of three or four classmates, study the following situations. Eliminate the negatives and find a fair, more positive way to talk about each situation.

Team Members

1. _____   2. _____

3. _____   4. _____

Situation 1: You're at lunch with a couple of teammates who are friends. There are a couple of new trainees in the crew, both of whom are having a hard time with some simple tasks. You sit around with your teammates and complain about how inept the two new co-workers are. Their ineptitude, you and your friends conclude, is holding everyone back, so you decide to freeze the new trainees out of any workplace conversation, unless you have to talk to them because the boss says so.

Situation 2: You are waiting on your co-worker, who's running late. You two were scheduled to start working on something first thing, but you can't do anything until the co-worker arrives. Meanwhile, you have a busy day, and every moment spent waiting increases the likelihood you'll have to stay late. Your co-worker never runs late, and you're wondering, why today? The only answer, you think, is that your co-worker has it in for you somehow. When your co-worker finally shows up, before the person has a chance to apologize you say, "Of all the days you pick to run late, today is that day. No accident it's the day you're working with me, is it? You know how much I have to do today, you knew it when you woke up this morning. So what do you do? Hit the snooze bar a few extra times. Then stroll in here like everything's peachy!"

Situation 3: You're delivering a load of sheet metal to another site. You're driving and your co-worker is supposed to be giving you directions. Instead, the co-worker gets involved in a cell phone call with the boss, and consequently you end up lost. You say to the co-worker, "I thought you were supposed to be on that map! I don't know where the heck we are now! We're lost and not going to get this stuff to the site on time, because you were chatting away on the phone! I don't care if it was the boss, you should have cleared that up before we left! I was counting on you for directions. Of course you probably can't read a map worth a lick anyway."

Activity 6: Passing the Criticism Buck

Many of us feel uncomfortable giving criticism. Many of us give criticism only when we're upset, so when we are being critical, we're often not in the best of moods. This is one reason why there tends to be more destructive criticism than constructive criticism. Work with three or four of your classmates on this activity. Choose one member of the team to be the scorekeeper and one member to be the timekeeper. Note that when these team members take a turn, they must hand over their duties to another team member. Keep in mind that your goal is to give constructive criticism.

Team Members

1. _____   2. _____

3. _____   4. _____

Rating Grid and Scorecard

Rate Points	Description
1	My team member did not do a very good job of giving constructive criticism. (Give suggestions for improvement.)
2	My team member did a pretty good job, but there is room for improvement. (Give suggestions for improvement.)
3	My team member did a good job. (Give examples of statements that you thought were helpful and constructive.)

Scorecard			
Team Member Name	Points for Giving Constructive Criticism (2)	Rating Points (from above)	Total
1.			
2.			
3.			
4.			

Materials Required

- Criticism slips (supplied by your instructor)
- Rating and scorecard (included)
- Watch or stopwatch
- Pencil

Your instructor will give your team several slips of paper. Each slip contains a situation for which constructive criticism must be given. To carry out this activity, follow these steps:

Step 1: Choose a criticism slip and read it. You must decide whether you want to use this information to give constructive criticism to your classmate or pass the slip on. Note that each of you must give at least one criticism before the activity ends.

Step 2: Say whether you will pass (and get no points) or go (for two points). If you decide to go, you have 30 seconds to think up what to say. If you can't figure out what to say in 30 seconds, you forfeit your turn and get no points.

Step 3: If you decide to go, turn to the teammate on your right or left and deliver the criticism.

The scorekeeper will keep track of points using the following chart. Although one of you may get more points than anyone else does, everyone who learns how to give constructive criticism wins in this game.

Activity 7: You, The Boss, or Forget About It

Earlier, we asked you to fill out a chart listing a number of workplace-conflict situations, and asked you to determine whether you should deal with the situation, whether your boss should, or whether you should just forget about the incident. In this activity, you will do something similar. Only this time, you will decide who should give the criticism, you or the boss, or whether you should just forget about it. After you have filled out the chart, discuss your answers with your team members. Did you all agree?

Team Members

1. _____ 2. _____

3. _____ 4. _____

	Situation	You	The Boss	Forget About It
1.	Your co-worker likes to sing in Spanish occasionally.	○	○	○
2.	Your co-worker uses unsharpened cutting tools.	○	○	○
3.	Your co-worker sneaks out 15 minutes early every Friday.	○	○	○
4.	Your co-worker misreads blueprints.	○	○	○
5.	Your co-worker's hair hangs loosely to the waist.	○	○	○
6.	Your co-worker borrows your tools and does not clean them.	○	○	○
7.	Your co-worker calls everyone "ace" and wears T-shirts with pictures of military vehicles like tanks on them.	○	○	○
8.	Your co-worker dozes during safety meetings.	○	○	○
9.	Your co-worker tends to hang back whenever there is hard work to be done.	○	○	○
10.	Your co-worker gives you praise but always adds a sarcastic comment.	○	○	○

Activity 8: Challenging Unfair Criticism

As you learned in this module, sometimes the person criticizing you is wrong. However, getting others to recognize that they may have made a mistake in criticizing you takes patience and skill. In this activity, you will work with three or four of your classmates to practice how to handle unfair criticism.

To complete this activity, choose one situation. Each team member must take a different situation. Take a few moments to figure out what you will say and how you will say it. As you prepare your remarks, keep the following in mind:

- Show respect at all times.

- State your case clearly and calmly.

- Don't make threats or become defensive.

As you state your case, your team members must listen carefully. Then discuss among yourselves how you each handled your situation.

Response Rating Grid

Response Rating Grid			
Observations	Yes	No	Tips for Improvement
1.			
2.			
3.			
4.			

	Observations	Yes	No	Tips for Improvement
1.	My team member was respectful.	○	○	
2.	My team members comments were clear and to the point.	○	○	
3.	My team member was calm.	○	○	
4.	My team member was defensive.	○	○	

Situation 1: Early this morning you got a call from T.J., who was supposed to pick up supplies on the way to work. T.J.'s truck broke down, so you volunteer to drop T.J. off at work and then go get the supplies yourself. As a result, you are late punching in and your boss reprimands you pretty harshly. You recognize that you should have told the boss about this situation earlier, but you also feel that because the boss does not have all the facts, the criticism is unfair.

Situation 2: You arrive at work early one morning to find a co-worker drunk and stumbling around the work site. You take your co-worker to a nearby clinic, and then report back to work. Unfortunately, your co-worker splashed some beer on you and you smell like you've been drinking. The boss takes one whiff, lectures you about your lack of personal responsibility, and says that you are fired.

Situation 3: Your boss reprimands you for measuring mistakes made in framing several rooms in a house. Yesterday, you questioned your boss about the measurements when you began the framing. The boss told you that the architect said the measurements were correct. You have to remind the boss about that.

Situation 4: Your co-worker suddenly becomes very cold toward you. You ask why and learn that your co-worker believes you are the source of nasty gossip now circulating around the site. Your co-worker criticizes you for being a gossip and spreading rumors. You have not gossiped and, in fact, were not even aware that gossip was circulating.

Sexual Harassment

"Do unto others as you would have others
  do unto you."

– The Golden Rule

"Do not do to others as you would not have them
  do to you."

– The Silver Rule

Introduction

In this module, you will learn how to treat all co-workers professionally and respectfully, regardless of gender or sexual inclination. You will learn how federal law defines sexual harassment in the workplace; the types of behavior that could be interpreted as sexual harassment; how to avoid sexually harassing your co-workers; and what you can do if you have been sexually harassed.

Sexual harassment at work occurs whenever unwelcome conduct based on gender or sexual orientation affects a person's job. Sexual harassment takes place in every type of job and in every type of industry. Many people believe that sexual harassment involves only men harassing women, and most lawsuits brought against companies do involve sexual harassment of women by men. However, costly lawsuits have also been brought against companies for both reverse harassment—women harassing men—and same-sex harassment.

Many workers are confused because some behaviors considered to be sexual harassment do not seem that serious. Telling jokes is often thought of as harmless fun; however, some types of jokes can be considered sexual harassment, depending on who hears them.

Sexual harassment is a serious issue. A company can be sued by workers who have been sexually harassed. Those lawsuits are costly, even if the case of sexual harassment is not proven. If it is, the company, as well as the individual employees who participated in the sexually harassing behavior, can be sued by the victim for damages, which can be in the thousands or even millions of dollars. Therefore, many companies have strict polices against sexual harassment and will fire workers who engage in such behavior. If you sexually harass a co-worker, it can have severe personal repercussions: you can lose your job, and it can damage your reputation in the industry.

Have You Experienced Any of These Situations?

- Have you heard obscene jokes on the jobsite? Were you ever the target of an obscene joke?

- Has a co-worker constantly been after you for a date or sex even after you've said no?

- Has a co-worker been angry with you because you refused to date or have sex with them?

- Have co-workers discussed their sex lives around you? Were you offended? Embarrassed?

- Have you overheard comments about a co-worker's physical characteristics? Have you heard such comments about your physical characteristics?

- Have co-workers called you insulting names that were also sexually suggestive?

- Have you been shown pornography at work? Was it offensive to you?

- Do co-workers stand so close to you that you get nervous? What if these co-workers were the same gender as you? How would you feel if they touched you inappropriately?

- Have co-workers threatened or harassed you because of your gender or sexual preference?

- Have co-workers suggested you are unfit for the job because of your gender?

- Have co-workers refused to work with you because of your gender or sexual preference?

Women in the workplace

Women make up a growing percentage of the construction workforce. According to U.S. Bureau of Labor Statistics data, in 2007 they comprised close to 33 percent of construction payrolls, and it is anticipated that the percentage of women in the construction workforce will increase. In spite of the success women are experiencing in the trades, sexual harassment against them persists. Several studies have shown that female construction workers experience widespread gender and sexual harassment. In one survey conducted by the National Institute for Occupational Safety and Health (NIOSH), during a one-year period, 41 percent of female construction workers experienced gender harassment. In another survey by the group Chicago Women In Trades, 88 percent of respondents reported being sexually harassed. In a NIOSH survey of some 500 female construction workers from across the nation, a majority of the respondents reported the following problems:

- They had to deal with a hostile work environment. They were made to feel uncomfortable by unwelcome suggestive looks, comments, joking, or gestures.

- They had to cope with nonexistent, inadequate, or unsanitary restroom facilities.

- They were reluctant to report sexual harassment incidents for fear of losing their jobs.

- They had been improperly touched or asked for sex.

- They had been issued personal protective equipment sized for the average man.

Definition of Sexual Harassment, and the Law

Sexual harassment is a form of sex discrimination that violates Title VII of the Civil Rights Act of 1964. The Equal Employment Opportunity Act of 1972 defines sexual harassment as "unwelcome sexual advances, requests for sexual favors, and other verbal or physical conduct of a sexual nature when:

- Submission to such conduct is made either explicitly or implicitly a condition of someone's employment, or

- Submission to or rejection of such conduct by an individual is used as a basis for employment decisions affecting that individual, or

- Such conduct has the purpose or effect of unreasonably interfering with an individual's work performance or creating a hostile, intimidating, or offensive working environment."

Courts and employers generally use these guidelines when dealing with cases or investigating claims of sexual harassment. A key word in the guidelines is *unwelcome*. Sexual conduct is considered unwelcome whenever the person subjected to it considers it unwelcome. In 1998, the U.S. Supreme Court ruled that employees in the workplace are to be protected from harassment by people of the same gender and that an employee does not have to suffer a tangible job detriment to sue for workplace sexual harassment. The Supreme Court has also established two legal definitions of sexual harassment in the workplace:

1. Quid Pro Quo Sexual Harassment: This describes a coercive situation when a job benefit is directly tied to an employee submitting to unwanted sexual advances. This would occur when a boss tells a subordinate worker that he or she will be fired if that subordinate does not cooperate sexually with the boss. This harassment is equally unlawful whether the victim resists and suffers the threatened consequences (in this case, being fired), or whether the victim submits to avoid those consequences.

2. Hostile Environment Sexual Harassment: A hostile environment is created when an employee is subjected to gender-based or sexually oriented unwelcome behaviors, such as comments about physical attributes, inappropriate touching, exposure to offensive sexual materials, or discussions of sexual activity.

Title VII is enforced by the Equal Employment Opportunity Commission (EEOC). When the EEOC investigates an allegation of sexual harassment, it looks at the nature of the sexual advances and the context in which those advances occurred. Here's an example of how the EEOC operates.

For example, on April 1, 2009, the EEOC filed a lawsuit in U.S. District Court against a road construction company, charging that the company violated federal law by subjecting a class of female employees to continual sexual harassment, and by retaliating against a woman who complained about the conduct, forcing the woman to resign. The lawsuit alleged that a boss made numerous sexual comments to women employees who were working a highway project, creating a hostile work environment. The lawsuit was filed after the EEOC had tried to reach a settlement with the company.

What is sexual harassment?

According to the EEOC, sexual harassment can occur in a variety of circumstances. These circumstances include, but are not limited to, the following:

- The victim or the harasser may be a woman or a man. The victim does not have to be of the opposite sex.

- The harasser can be the victim's supervisor, an agent of the employer, a supervisor in another area, a co-worker, or someone not employed by the company (for instance, someone employed by a supplier or subcontractor).

- The victim does not have to be the person harassed but could be anyone affected by the offensive conduct.

For unlawful sexual harassment to occur, it does not require economic injury to the victim nor does it require a threat to the victim's job.

Only unwelcome conduct can be considered sexual harassment; if the joking or touching (or dating) between co-workers is welcomed by both of them, this is not considered to be sexual harassment. Conduct is considered unwelcome if the recipient did not initiate it and regards the conduct as offensive. The following behaviors are considered to be sexually harassing at the workplace if they are unwelcome:

Sexual advances or requests for sexual favors. Sexual advances means more than just asking for sex. It also means pestering a co-worker for a date when that co-worker has said no.

The clearest case of unwelcome conduct would be when the employee directly tells the harasser that the sexual advance or other sexually oriented conduct is unwelcome, that the behavior is making the employee uncomfortable, and that the harasser must stop the behavior immediately.

A more ambiguous case is when the offended employee has not been clear about whether the conduct is considered unwelcome. For instance, if you ask a co-worker out on a romantic date and the co-worker mutters, "No, no, sorry, I have a previous commitment," your request probably would not be considered sexual harassment because your invitation wasn't fundamentally offensive and your co-worker's response was too vague to determine whether or not the co-worker considered your conduct unwelcome.

Another case involves co-workers who have been in a sexual relationship when one of them wants to end the relationship but the other does not and is insistent about it. If you have been in a sexual relationship with a co-worker and you want to break it off but your ex-partner does not, and if that co-worker's continuing sexual advances to you at work are affecting your job, then your ex-partner's behavior would be considered unwelcome and therefore sexually harassing. Employees have a right to end relationships with co-workers without fear of retaliation on the job.

Other verbal or physical conduct of a sexual nature. Sexual harassment can be either physical or verbal. Examples of physical conduct of a sexual nature include:

- Standing so close to co-workers that they feel uncomfortable. This can include leaning over co-workers, cornering them, or gratuitously hugging them.

- Touching a co-worker inappropriately or touching a co-worker who does not want to be touched. This can include pinching and unwanted neck or back massages.

- Making obscene or vulgar gestures, such as smacking lips or touching oneself in a sexual manner.

Examples of verbal conduct of a sexual nature include:

- Making vulgar comments or sounds (such as whistling, howling, kissing sounds, or cat calls).

- Referring to an adult using such words as *girl, doll, babe, honey, hunk,* or *stud.*

- Sending unwanted letters, telephone calls, emails, or text messages of a sexual nature.

- Calling co-workers by sexually offensive names.

Making sex or sexual favors a term or condition of employment. Demanding sex from someone in exchange for keeping a job or getting a promotion is an example of *quid pro quo* sexual harassment. The definition of *term or condition of employment* is not just limited to a boss-worker situation, however. If a worker takes negative action against a co-worker or threatens that co-worker for refusing sexual advances, such behavior is considered sexual harassment even if the harasser is not the victim's boss.

Creating an intimidating, hostile, or offensive working environment. Sexual harassment is not limited to requests or demands for sexual favors. It can also be any gender-based offensive behavior that makes the workplace threatening or offensive. In a sexually hostile environment, unwelcome conduct is abusive to the victim. It is best to prevent a sexually hostile environment from being established by avoiding these behaviors in the first place. Some examples of behaviors that create an intimidating working environment are:

- Telling obscene jokes.

- Engaging in suggestive sexual talk or sexual innuendo.

- Asking about a co-worker's sexual fantasies, preferences, or histories.

- Telling lies or spreading gossip about a co-worker's sex life.

- Displaying pornographic calendars, cartoons, or pictures.

- Using email to send pornographic pictures, stories, or jokes.

- Using profanity.

- Referring to co-workers using demeaning names or terms.

- Staring at a co-worker's body or making inappropriate comments about a co-worker's physical attributes.

- Stating or implying that co-workers are less competent because of their gender.

- Making personal comments about a co-worker's social or sex life.

- Turning a work discussion into one about sex or sexual topics.

In most cases, sexual harassment involves more than one isolated incident; generally it is considered a pattern of behavior repeated over time. However, sexual harassment can be based on only one incident, and there have been a number of sexual harassment legal actions taken as a result of a single event.

How to avoid sexually harassing others

Often, people accused of sexual harassment honestly do not understand the accusation. They may feel they did not mean any harm, they are not sure what they did wrong, and they are surprised that they're being accused of sexual harassment.

The best way to avoid being charged with workplace sexual harassment is to think carefully about what you do and say around co-workers. You may have to change your behavior or your conversational style. Here are some tips that will help you to avoid sexually harassing your co-workers.

Be a professional. Remember the qualities of being a professional worker. At all times on the job and with everyone, act that way.

Professionals behave with dignity and treat others with respect. Unfortunately, construction workers are sometimes portrayed as loud, rude people who treat others, especially women, badly. Do not let your behavior add to this unfair stereotype.

Don't comment on anyone's physical appearance. You never know how such comments will be received. If you compliment a co-worker's hair, figure, or clothing, that person may be flattered—or that person may be annoyed and wonder what your real intention is. When you make a comment about a co-worker's physical appearance—even if you mean it simply as a compliment—you are stepping onto uncertain ground. If you want to compliment co-workers, by all means do so but stick to work-related characteristics: the good job they do, the positive attitude they bring to team, the consideration they show fellow workers.

Don't use terms of endearment. You may not mean any harm when you call another co-worker *honey* or *sweetie*, but you don't know how these terms will be received.

Think before you speak. Are the people you work with comfortable hearing dirty jokes? Do you think that everyone in the group really wants to hear about your sexual adventures? The best course of action is to avoid off-color jokes and keep quiet about sexually oriented topics.

No means no. Do not assume that a co-worker you are trying to date is playing hard to get if they do not want to go out or have sex with you. If a co-worker won't date you or refuses your advances, then you must stop making those advances and drop the entire issue, or face being subject to a sexual harassment lawsuit.

Remember the empathy rule. How would you feel if you worked in a hostile environment where everyone prejudged you, treated you disrespectfully, or harassed you because of your gender or sexual orientation? Even if you don't find a conduct personally offensive, remember that some of your co-workers might. If you are uncertain whether your behavior is unwelcome, ask yourself these questions: Would I change my behavior if a family member were here? Would I want a member of my family to be treated this way?

Read your company's sexual harassment policy. Most companies have a policy regarding sexual harassment. Read your company's policy and make sure you understand it.

Watch the use of email and text messaging. Don't send out emails from work with suggestive or potentially offensive material in them, even if you're only sending it to people you know won't be offended. Someone might be looking over the recipient's shoulder. The same rules apply to texting.

What to do if you've been sexually harassed

You have the right to work in a nonthreatening environment. You do not have to suffer in silence if you feel you are a victim of sexual harassment. These are some of the measures you can take in response:

- Confront the person who is sexually harassing you. You can say something like, "I don't like sexually oriented jokes" or "Take your hands off me right now! I won't put up with you touching me." Some experts also recommend writing a letter to the harasser as a way of showing how serious you are. In any event, your first move is to tell the harasser clearly and directly to stop or otherwise you will take the issue to their supervisor.

- Keep a written record of the incidents, including what exactly happened, when it happened, and the names of those who either participated or witnessed the harassing behavior.

- File a complaint. If the harasser(s) continue their behavior, you can complain to your supervisor or their supervisor. This is usually the best plan if directly confronting the harasser does not work. When you make your complaint, provide the specifics of time, place, and what was said or done. (That is why you should keep a written record of the incidents. When you make a complaint of this sort, it is always best to document it.)

- Remember, what's important is that you have been sexually harassed at work. The harasser does not have to be an employee of the company.

There may be times when you face a difficult circumstance; for example, when you tell your boss you are being sexually harassed and nothing is done about it. There is the possibility you could face an even more unpleasant situation: when it is your boss who is the harasser and you are worried that if you do not submit to your boss's advances, you will lose your job. You might think you are in a real bind if you find yourself in either of these situations, but you can and should take action. Your boss's status as your supervisor should not intimidate you from doing what you think you need to do to stop the sexual harassment. In these situations, you can take the following measures:

- First, consult your company's employee handbook. The handbook should include your company's policies regarding sexual harassment, including a chain of command you should follow to report any incident, and other specifics regarding the company's grievance or complaint procedures.

- Next, follow that chain of command and bring the incident to the attention of the appropriate individual in your company.

- If you file a complaint through your company's complaint process, make sure it is written.

- If you still do not get any satisfaction, or if the company you work for does not have procedures for dealing with sexual harassment, then you may have to consult with a sexual harassment attorney as a last resort. Many attorneys provide free consultations about these situations.

Summary

Sexual harassment at work occurs whenever unwelcome conduct based on gender or sexual orientation affects a person's job. The victim or the harasser may be a woman or a man. A victim of sexual harassment could be anyone who is offended by offensive conduct, not just the person being harassed. The harasser can be the victim's supervisor or co-worker, or someone not employed by the company. Sexual harassment comes in a number of forms: sexual advances, requests or demands for sexual favors, physical conduct such as inappropriate touching or obscene gestures, and verbal conduct such as vulgar jokes or sexually offensive references. That means you put your job and career at risk if you engage in it, and that also means you do not have to put up with it. To avoid sexually harassing your co-workers, act like a professional on the job at all times and treat everyone with respect. Avoid commenting on your co-workers' physical appearance, think before you speak or act, and know what your company's sexual harassment policy is. If you feel you've been a victim of sexual harassment, confront the harasser first and tell that person (or persons) to stop the behavior. If that does not work, record the details of the incident, when it occurred, and who was involved. Report the issue to the harasser's supervisor. If the supervisor does nothing about it, or if your boss is sexually harassing you, follow your company's procedures to file a complaint.

Individual Activities

Activity 1: Understanding When Sexual Harassment Occurs

Several situations are listed below. Your task is to determine whether or not sexual harassment has occurred in each case.

Is This Workplace Sexual Harassment?	Yes	No
1. Linda reports to John, her supervisor. For the past several days, Linda has punched in late. John gently takes Linda's arm to pull her off to the side so that he can talk to her in private, removes his hand, and then quietly but sternly tells her she must report to work on time or face the consequences.		
Did John sexually harass Linda?	○	○
What if John grabbed Linda roughly by the arm, but said or did nothing else different—would this be sexual harassment?	○	○
2. Co-workers Fran and Steve are attracted to each other. They start to date and, when no one is around, Steve calls Fran "sweetie" and Fran calls Steve "baby."		
Did Steve sexually harass Fran?	○	○
Did Fran sexually harass Steve?	○	○
3. Co-workers Brenda and Richard dated for a while and then broke up. Richard wants to get back together, but Brenda does not. Richard starts hiding Brenda's tools, has intentionally spilled his soda on Brenda twice, and has been heard muttering abusive words when she walks past.		
Is Richard sexually harassing Brenda?	○	○
4. Ellen is interested in Ken. She always smiles at him and asks how he is doing. Today, Ken told Ellen how nice she looked and asked her if she'd like to have lunch with him. Ellen agrees to have lunch with Ken and says she is glad he asked.		
Did Ken sexually harass Ellen?	○	○
5. Several male workers post pictures of naked women in the break room, so some of the women put up pictures of naked men.		
Were the men sexually harassing the women?	○	○
Were the women sexually harassing the men?	○	○

Is This Workplace Sexual Harassment?   Yes   No

6. At your company, workers use email to communicate information about tasks and schedules. One worker starts using the system to send pictures of models in skimpy swimsuits.

   Was anyone sexually harassed?   ○   ○

   If a worker complained about the pictures, would this be a case of sexual harassment?   ○   ○

   If the pictures were pornographic, but no one has complained, would this be a case of sexual harassment?   ○   ○

7. One of your company's suppliers is pestering you for a date. Whenever the supplier is on site, you feel uncomfortable because you know you will be subjected to unwanted advances.

   Are you being sexually harassed?   ○   ○

8. Your co-worker is being sexually harassed by another co-worker. You are not directly involved, but the daily confrontations between these two are affecting your ability to do your job. In addition, you find the harasser's conduct offensive even though it is not directed toward you.

   Are you being sexually harassed?   ○   ○

Substance Abuse on the Job

"If a person has stable housing but has a current… substance abuse problem with which they are not getting any support, it is going to put a strain on [their] employment."

– Adam Sampson,
author and social scientist

Introduction

The term *substance abuse* does not have an absolute definition. Generally, *substance* refers to alcohol and drugs—both illegal (marijuana, ecstasy, and crystal meth) and prescription (Ritalin and Oxycontin)—that modify mood or behavior. However, some substances that are not technically drugs, such as glue, can be abused for the effect they have, and some drugs that can be abused, such as steroids, do not produce a high (although they can alter a person's mood). Even caffeine, which is a mild stimulant, can be used excessively—as anyone who's had a few cups of coffee on an empty stomach can tell you.

Many companies have adopted zero-tolerance policies on the use of illegal drugs at any time and place, and on the use of alcohol on the job. If you have a problem with alcohol or prescription medications, if you use illegal drugs, or if you drink any alcohol before or during work, it could cost you your job or prevent you from getting one in the first place. Most important, any use of drugs or alcohol before or during work is a real safety hazard; it could lead to your death or the death of your co-workers. Any use of alcohol or drugs on the job is extremely dangerous, which is why it is strictly forbidden.

Substance abuse and alcoholism can lead to overdoses and death; they are not problems that go away on their own. However, they are treatable. This module will discuss substance abuse, its effects and dangers, and how to recognize it and deal with it in others—or even in yourself.

Substance Abuse...

...means different things to different people. People who use such illegal drugs as heroin, crack, or crystal meth are clearly substance abusers. People who are addicted to or dependent on any drug (including alcohol) are substance abusers as well. But what about the person who rarely drinks, but one night really ties one on? What about someone who smokes the occasional joint, but only on the weekend or after work? Many people think that substance abuse includes *any* use of *any* illegal drugs and that they are all dangerous; others argue that the occasional and casual use of marijuana is no more harmful than drinking or smoking tobacco. The fact remains that whether you think marijuana is no worse than booze, its use is illegal; therefore, if you test positive for it, even if you only smoke the occasional joint on the weekend, you can be fired from your job, no questions asked.

Substance abuse among construction workers

The rates of substance abuse among workers in construction are among the highest in any industry. According to a 2007 study by the U.S. Government's Substance Abuse and Mental Health Services Administration (which is part of the Department of Health and Human Services),

- 17.8 percent of those in the construction industry reported heavy alcohol use in the past month;
- 16.9 percent of workers reported alcohol dependence or abuse during the past year;
- 6.2 percent reported dependence or abuse of illicit drugs during the past year;

These workers are more likely to have worked for more than three employers in the past year, and are more likely to have skipped work for more than two days in the past month.

Substance abuse by workers is an obvious safety concern and can lead to problems with overall worker morale:

- Up to 32 percent of workers in all industries have had their job performance affected at one time or another by a co-worker's drug or alcohol use;

- Up to 40 percent of industrial fatalities and 47 percent of industrial accidents can be linked to alcohol abuse and alcoholism;

- The losses to all companies in the United States due to alcohol- and drug-related abuse by employees could be more than $100 billion a year, directly and indirectly.

It has been estimated that six out of ten adults know someone who has reported to work while under the influence of drugs or alcohol. Studies suggest substance abusers are two to four times as likely to have an accident on the job as those who do not abuse substances, and substance abusers can be linked to about 40 percent of all work-related deaths in the United States. Between 10–20 percent of those who die in work-related incidents have alcohol and/or drugs in their system. The highest rates of drug use on the job are found in industries (one of which is construction) where the chance of being injured or killed on the job is the greatest.

Substance abusers do not have to drink or take drugs on the job to have a negative effect on the workplace. Depending on how the term is defined, substance abusers, compared with their co-workers who are not substance abusers, are:

- Up to 10 times more likely to miss work;

- More than three times more likely to be involved in an on-the-job accident;

- More than five times more likely to be injured at work; and

- Less productive; studies have shown that substance-abusing workers function at only about two-thirds of their capability.

The effects of substance abuse

Substance abuse can cause serious problems—health, legal, personal—for the individual, and can lead to serious problems for that person's family and employer. The use of illegal drugs and the abuse or irresponsible use of alcohol and prescribed drugs can lead to arrest and incarceration, and the abuse of any substance, whether legal, illegal, or prescribed, can lead to family problems, addiction, long-term health problems, and even death.

As you begin or continue your career in construction, remember the following points:

- **Substance abuse endangers your safety and that of your co-workers.** Construction work is often dangerous, involving the use of high-speed tools and equipment, potentially harmful solvents or other hazardous materials, and work in deep trenches or far above the ground. You must be on top of your game at all times, agile and quick thinking. Construction sites are full of potential hazards; therefore, even if you are not doing something that's dangerous, you have to be aware of your surroundings. Would you want to be working up high on a scaffold with a co-worker who was drunk? Or on a scaffold put together by a worker under the influence? How safe would you feel if a co-worker under the influence was operating a crane nearby? Workers with substance abuse problems, even if they do not indulge before or during work, can still present a safety hazard because prolonged substance abuse can dull reflexes and affect judgment.

- **Substance abuse or use of illegal drugs can prevent you from getting the job you want.** As we have seen in numerous places in this Workbook, safety is absolutely vital to construction work. Therefore, many employers screen applicants for drugs and alcohol. If your tests show that you take illegal drugs or that you abuse alcohol or prescription drugs, even on days when you are not working, you will not get the job.

- **Substance abuse or use of illegal drugs can get you fired on the spot.** Because safety is so critical to construction work, most construction companies have a zero-tolerance policy for workers who drink or take drugs on the job. This means you will be fired immediately. Many companies also have a zero-tolerance policy on taking any illegal drug, any time, and they do random drug testing of their employees; if you test positive, you will probably be fired.

- **Substance abuse damages your well-being and quality of life.** Substance abusers do not perform at their best. They often have trouble paying their bills, dealing with emotional problems, and managing their lives. Drugs and alcohol keep them from thinking straight, and they tend to lie, make excuses for their behavior, or blame others for their problems. Substance abuse, particularly if it is prolonged, can cause real damage to your mental skills and your motor skills (your ability to use tools and equipment)—skills that are critical to doing a good and safe job as a construction professional.

- **Substance abuse jeopardizes your relationships.** Substance abuse can do great damage to your ability to get along with your friends, family, and co-workers. Consider how substance abuse affects your personal relationships:

  - Your family cannot depend on you to work a steady job or to provide emotional and financial support if you are spending all your money on alcohol or drugs.

  - Your friends cannot depend on you to keep promises or pay back loans.

  - Your co-workers cannot depend on you to do your share of the work, and they cannot trust you to work safely.

What About That Lunchtime Beer?

It's a hot day, and the idea of one cold beer at lunch can be tempting. You don't have any problem with alcohol, and you're highly regarded by your boss and co-workers. So you're thinking—a cold one couldn't hurt. Well, yes it could. Companies generally do not want their employees drinking on the job at all. Even one beer can affect your judgment, balance, and reflexes, qualities you need to work safely at a jobsite. Some companies have been known to do random breathalyzer testing of their employees on the job, and even a little bit of alcohol will show up on those tests. Don't have that lunchtime beer; it could cost you your job.

Prescription drugs

When people talk about substance abusers, they usually mean alcoholics or people who take illegal drugs. However, many people also abuse prescription drugs, sometimes on purpose and sometimes without realizing it. In fact, a 2008 report from the drug-testing company Quest Diagnostics found that more workers are now testing positive for prescription drugs than for cocaine and amphetamines combined. Because prescription drugs are legal and prescribed by a doctor, many people do not think of them as harmful. However, people can become addicted to certain types of prescription drugs—painkillers, sedatives and tranquilizers, and stimulants. Abuse of prescription drugs can have harmful effects similar to those from abusing other drugs and alcohol. It can dull your senses, your reflexes, and your judgment, and it can make you a threat to both your own safety and that of your co-workers on the job.

Borrowing Someone Else's Medication

What if you need a prescription—for instance, you've got an injury that's slowing you down at work and you need some pain pills—but you don't have health insurance and don't think you can afford to see a doctor to get a prescription? The temptation is to find a willing co-worker who's taking some medication and borrow a pill from that person every now and then. You should not do this, because it is against the law to transfer prescription medications. It is also potentially dangerous to take them while not under a doctor's supervision.

If you are taking prescription drugs, keep the following things in mind:

- **Take prescription drugs only as directed.** Your doctor and pharmacist will give you information about the drug and how and when it should be used. They will also tell you about side effects and what not to take or do while taking the drug. Before you get a prescription, make sure the doctor who gives it to you knows that you work in construction.

- **Understand the side effects.** Some drugs can make you drowsy or less able to concentrate. You must tell your supervisor if you are taking a prescription drug that could affect your motor or mental skills. You must not drive or operate power tools or heavy equipment when taking certain types of prescription drugs. Your doctor will be able to tell you which drugs you should tell your supervisor you are taking.

- **Do not extend or share your prescription.** When your prescription period is up, you must return to your doctor for an examination to get more medication if you need it. Do not lie to your doctor to get more medication; if the symptoms have stopped, quit taking the medication. Do not talk friends into getting your prescriptions filled for you. Never share your prescription medication with anyone else.

- **Know when to ask for help.** If you find that you cannot get through the day without a prescription drug, see your doctor immediately. People can become addicted to prescription drugs, especially painkillers, sometimes without realizing what is happening. Your doctor can help you reduce your dependence on the drug safely.

A Word About Steroids

Yes, they are a drug; yes, they are dangerous; and yes, if you are taking them without a prescription, you are breaking the law. Steroids cause violent mood swings and in the long run increase the risk of heart disease, stroke, and cancer. Steroid users develop acne and lose their hair. In men, steroid use causes testicle shrinkage and impotence; in women, it causes development of male traits such as a deeper voice and facial hair.

Recognizing and coping with substance abuse

At some point in your career, you may find yourself working with someone who has a substance abuse problem. Sometimes, it will be easy to spot—a co-worker may show up smelling of alcohol, stagger, or have slurred speech. Seeing someone drink alcohol or take drugs on the job is an obvious sign of trouble, and if the co-worker is so blatant about it that he or she evidently does not care whether co-workers see it, that is a strong sign of a substance abuse problem. At other times, however, the signs of substance abuse are not as noticeable.

Your supervisor will probably notice a substance abuse problem before you do. If your boss is offsite a lot while the problem is developing, however, then you (or another worker) might notice it first. As a professional member of the work team, you must play your part in making sure that an impaired worker does not endanger others. In other modules, you learned that being a tattletale is not a good idea; this is one exception to that rule. You must report the behavior of an impaired worker to your boss immediately. Your boss can then step in to ensure that nobody gets hurt. If you see a co-worker who is under the influence, it is not just that employee's problem; it is a problem for the entire work crew.

The signs of substance abuse

The following list includes many of the signs of substance abuse. It is not your responsibility to figure out why a co-worker is having a problem or what substance the co-worker is abusing. You simply need to be aware of behaviors that indicate a co-worker may have a problem.

You should be aware that not everyone reacts the same way to a specific substance. Some of these signs are not necessarily indications of substance abuse; they could be indications of another health problem. Any one of these warning signs in isolation should not be considered a sign of substance abuse; there are a lot of other reasons why a co-worker might be always late or always moody. Bloodshot eyes might just be a sign of a lack of sleep the night before. The more of these signs you see, however, the greater the chances that your co-worker has a problem and needs help.

The Warning Signs of Substance Abuse

- Often arrives late to work
- Often leaves work early
- Takes days off for unexplained reasons
- Sometimes disappears from the jobsite during the day for prolonged periods
- Suffers frequent colds, headaches, or other ailments
- Often disappears from the jobsite for short periods
- Does not show up for work at all
- Seems disconnected from what's going on
- Loses interest in grooming or starts showing indifference to personal hygiene
- Loses a lot of weight suddenly
  (or in the case of steroids, gains a lot of muscle all of a sudden)
- Acts moody or paranoid, or otherwise unpredictably
- Reacts oddly or violently, particularly to common work situations
- Borrows money frequently
- Slurs speech, walks unsteadily, or has bloodshot eyes
- Seems unusually tired or unusually hyper
- Has serious financial or family problems

Effect on Job Performance

- Misses deadlines often
- Makes more mistakes than usual
- Does poorer quality work than usual
- Is not as productive as usual
- Leaves co-workers to pick up slack
- Lies to cover up absences and mistakes
- Doesn't pay attention or can't concentrate
- Works erratically
- Works more slowly (or much more quickly) than usual
- Often involved in accidents both on and off the job
- Often cited for safety violations at work
- Sustains a series of minor injuries
- Injures co-workers
- Shows anger or hostility toward co-workers and supervisors
- Has escalating problems with co-workers and supervisors
- Picks fights and has arguments at work

Do you have a substance abuse problem?

If you have been reading this module and think you have, or could be developing, a substance abuse problem—get help. For one thing, you will lose your job if you do not (or you will not have any success in finding one). If you are caught using drugs or alcohol on the job but you can prove you are in a substance abuse program, you may be less likely to be fired on the spot.

To get help, contact any of the organizations listed in *Appendix A*. At least one of these groups should have an office in your city or town or in a nearby city. Local offices for many of these organizations are listed in your telephone book or online. In addition, a hospital or clinic in your town may offer programs that can help you.

Read your employee handbook. It probably has a section on employee substance abuse, and information on company assistance for employees who admit to a substance abuse problem before being caught. Your company will be a lot more understanding if you admit the problem before it negatively affects your job performance, or before the problem makes you a safety risk. Don't allow a substance-abuse issue to escalate until it becomes a serious work-related issue; at that point, your company will probably show you little or no sympathy.

Summary

If a worker drinks or takes drugs on the job, or if a worker has a problem with drugs or alcohol, that person is endangering the safety of everyone on the jobsite. Because construction work is so demanding and potentially dangerous, and because workers need to be aware and alert at all times, many companies have zero-tolerance policies on the use of drugs and alcohol on the job, and many have a zero-tolerance policy against their employees using illegal drugs at all. Substance abuse—not only of illegal drugs but also of alcohol and prescription medications—can wreck lives as well as careers, and can lead to long-term health problems, and even death. If you think a co-worker has a substance abuse problem or is using drugs or alcohol on the job, you need to report it to your supervisor. Therefore, you should know the warning signs of substance abuse. If you believe you have a substance abuse problem, get help immediately.

On-the-Job Quiz

Here's a quick quiz that allows you to apply what you've learned in this module. Select the best possible answer, given what you've learned.

1. You are nervous about an interview for a construction job, so you grab a few beers before the interview to calm down. What do you think is most likely to happen?

   a. Your interviewer will understand that nerves made you drink, but will ask you if you have a drinking problem.

   b. It's almost certain you will not be hired.

   c. You will be told to sober up and come back to try again.

   d. No one will notice because you really did not have that much to drink.

2. At a job interview, the interviewer asks whether you take drugs. You smoke marijuana occasionally but figure that doesn't count, so you say, "No." You then take a drug test, and it detects marijuana in your system. What do you think is most likely to happen?

   a. The test won't matter because you smoked the marijuana before you were hired.

   b. The test will detect such a low level that it will not matter.

   c. The test will detect the marijuana, the positive test will be compared to your denial, the company will decide that you lie and cannot be trusted, and you won't be hired.

   d. You will be called in for a second interview, where you'll be lectured about lying and marijuana smoking, but will be hired anyway on a probationary basis.

3. You drink all night on Sunday, grab an hour or two of sleep, have a couple of drinks first thing Monday morning, and show up to work smelling of alcohol and swaying back and forth. Your boss will probably _____.

   a. reprimand you

   b. fire you on the spot

   c. lecture you about the evils of alcohol

   d. discipline you by suspending you for two weeks without pay

4. You and a co-worker always get lunch off the jobsite at a bar, where you eat sandwiches and fries and have a couple of beers each. Is drinking a couple of beers at lunchtime acceptable?

   a. Yes. A couple of beers won't affect job performance.

   b. It is not acceptable, but it isn't really unacceptable, either, given that the food absorbs the alcohol and you will not be affected by it.

   c. It is not acceptable because you shouldn't drink any alcohol during the workday, even if you think it doesn't affect you.

   d. It's probably acceptable because you've never heard anyone say you could not have a beer at lunch during the workday.

5. You and a co-worker are up on a scaffold to nail siding using pneumatic power nailers. You notice that your co-worker smells of beer. Once up on the scaffold, your co-worker looks around, sways a little, removes a flask from a hiding place, and takes a long swig. Should you be concerned?

    a. No. Your co-worker was able to climb the scaffold and so should be able to operate the power nailer safely.

    b. Yes. Your co-worker might be drunk or not, but the fact the co-worker smelled of beer, swayed around, and tried to hide the flask while drinking out of it are enough reasons to be concerned. It is your responsibility, then, to tell your boss these things, and let your boss decide what to do about it.

    c. Yes. You should tell your co-worker to knock off the drinking because this is a dangerous job.

    d. Yes, but you don't say anything because the flask might contain water, soda, or juice. If you tell the boss and it turns out the co-worker's flask didn't have alcohol in it, you'll look like a fool and a busybody.

6. You were under a lot of stress, so a doctor gave a 30-day prescription for Valium to calm you down. You're under a lot less stress now, feeling good, and working better than ever, and want to stay on the Valium without having to go back and pay for the doctor. You're able to get refills at the drug-store through a friend. A co-worker hears you are taking Valium and asks for one periodically, and you are glad to share because the co-worker is going through a tough time just like you were a while back, and you empathize. Are you doing anything wrong here and, if so, what?

    a. Yes. Your friend could get into serious trouble filling the prescription for you, even though you shouldn't have to go back to the doctor to get it refilled. It's no big deal giving one to your co-worker every now and then, however, because you help your teammates when asked.

    b. Yes, you should never have taken Valium in the first place; it is just a crutch for people who are weak-minded.

    c. Not really. You are kind of breaking the law, but no one is going to catch you, and Valium is a mild drug anyway, especially the dosages you take and give to your co-worker.

    d. Yes. You are breaking the law by abusing a prescription drug and giving it to people other than yourself.

7. You are doing a two-person job that involves operating some heavy equipment. Before you started, you saw your co-worker swallow a couple of pills with a soda. You think you smell alcohol as well, and you are wondering if there was anything other than soda in that can. What is your best course of action?

    a. Tell your co-worker you know what was in that can and to give you a drink; otherwise, you'll report them for drinking on the job.

    b. Ask what pills the person took with the soda, and then give a lecture about the dangers of taking pain medication while operating heavy equipment.

    c. Tell your boss that you think the co-worker may be drunk and has been swaying around. Explain that the person swallowed several pills before starting to operate the equipment (although you don't know what they were) and that you don't think this person is capable of operating potentially dangerous equipment in this condition.

    d. Gently ask the co-worker after the day is done whether the person has a substance abuse problem.

8. You've been working with T.J. for 3 years. On Mondays, you usually talk about what you did over the weekend. Today, T.J. tells you about a reunion with old friends at which everyone got drunk. This is the first time T.J. has ever described drinking or getting drunk on the weekend. What is your most reasonable assumption?

   a. T.J. probably doesn't have a drinking problem; T.J. just got drunk over the weekend, and that's not a sign of alcohol abuse in itself.

   b. T.J. is not used to drinking and may have driven home drunk. You should say something to T.J. about the dangers of drinking and driving.

   c. T.J. might be developing a serious alcohol problem because T.J. never drinks like that. You should keep an eye out.

   d. T.J. is probably an alcoholic, and you should either tell the boss immediately or suggest that T.J. get counseling.

9. Your co-worker K.C. has historically been one of the most dependable and good-natured members of the work crew, always showing up for work well-groomed and ready to go. Lately, however, K.C. often shows up in dirty clothes, is slow to get started in the morning, seems cranky and tired all the time, disappears from the worksite regularly, and then comes back using slurred speech. K.C.'s also been making a lot of (so far) minor mistakes. What is your best course of action?

   a. You suppose that K.C. could be drinking or doing drugs during the disappearances, but because it hasn't seriously affected the job yet, you figure it is not quite time to tell the boss, who has probably noticed these suspicious behaviors anyway.

   b. You confront K.C. about the disappearances and the changes in appearance and habits.

   c. Even though the boss may have noticed some of the changes in K.C's behavior, appearance, and attendance, you talk to the boss anyway because, taken together, these behaviors strongly suggest a possible substance abuse problem; if the boss had known about them all, there probably would have been action taken already.

   d. You give K.C. a pamphlet from Narcotics Anonymous and say that you know what's going on; threaten to let the boss know if K.C. doesn't get help.

10. You've tried everything, but you just can't gain weight and muscle mass. You start taking steroids you get from an acquaintance and you start bulking up. You get a lot of admiration from your co-workers, and the one person on the team who used to bully you has suddenly stopped. The other day, you got rough in an argument with a friend of yours at work, which is not like you at all. You were reprimanded by your boss, who was clearly a bit surprised at your behavior. What is your best course of action?

   a. Take something like Valium to keep you calm while you are on steroids, so that you do not get aggressive but still get the benefits.

   b. Lower your dosage of steroids.

   c. Quit taking steroids immediately because they are affecting your mood to the point that you might end up being fired for acting violently at work, and could be doing a lot more damage, too.

   d. Keep taking the steroids because you like all the positive attention, but decide to keep a better hold of your temper, make a special effort to be nice to your friend, and listen all the more carefully to what the boss says to help smooth things over with both of them.

Getting More Information About the Effects of Substance Abuse

Many organizations are devoted to helping people and families facing problems with alcohol or drugs. Contact any of the organizations listed here and ask them to send you a packet of information, or visit their websites. Some of these organizations may have an office in your town, which you can find by looking in your local telephone book. Many of these organizations provide toolkits, fact sheets, and activities that can help you prevent or deal with the problems of substance abuse. Share this information with the class or with co-workers.

Al-Anon/Alateen

Family Group Headquarters, Inc.

1600 Corporate Landing Parkway

Virginia Beach, VA 23454-5617

888-4AL-ANON

http://www.al-anon.org

Alcoholics Anonymous

World Services, Inc.

475 Riverside Drive

New York, NY 10115

212-870-3400

http://www.alcoholics-anonymous.org

Cocaine Anonymous

World Service Office

3740 Overland Avenue, Suite C

Los Angeles, CA 90034

800-347-8998

http://www.ca.org

Marijuana Anonymous

World Service Office

P.O. Box 2912

Van Nuys, CA 91404

800-766-6779

http://www.marijuana-anonymous.org

Mothers Against Drunk Driving (MADD)

511 E. John Carpenter Freeway, Suite 700

Irving, TX 75062

800-GET-MADD (438-6233)

http://www.madd.org

Narcotics Anonymous

World Service Office

P.O. Box 9999

Van Nuys, CA 91409

818-773-9999

http://www.na.org

National Association for Children of Alcoholics

11426 Rockville Pike, Suite 301

Rockville, MD 20852

301-468-0985

888-554-COAS

http://www.nacoa.net

National Clearinghouse for Alcohol and Drug Information

U.S. Department for Health and Human Services

P.O. Box 2345

Rockville, MD 20847-2345

301-468-2600

800-729-6686

http://www.health.org

National Council on Alcoholism and Drug Dependence

244 East 58th Street, Fourth Floor

New York, NY 10022

800-NCA-CALL

http://www.ncadd.org

National Highway Traffic Safety Administration

1200 New Jersey Ave., SE, West Building

Washington, DC 20590

1-888-327-4236

Auto Safety Hotline: 800-424-9393

http://www.nhtsa.dot.gov

National Institute on Alcohol Abuse and Alcoholism

5635 FIshers Lane, MSC 9304

Willco Building

Bethesda, MD 20892-9304

301-443-3860

http://www.niaaa.nih.gov

Facts About Alcohol

- A 12-ounce bottle of beer or wine cooler contains as much alcohol as one 5-ounce glass of wine or 1.5 ounces of 80-proof distilled spirits.

- In almost every state in the United States, you are legally drunk as soon as your blood alcohol level (the percentage of alcohol in your bloodstream) equals 0.08 percent.

- Someone in America dies every 30 minutes as a result of an alcohol-related vehicle crash.

- Alcohol abuse creates a reduced sensitivity to pain and an altered sense of time. For people who are drunk, time appears to pass more rapidly.

- Alcohol generally does not improve sexual performance.

- Binge drinking is not itself a sign of alcoholism, but it could be one of the signs.

- People most likely to drive under the influence of alcohol are between the ages of 21 and 34.

- On average, three out of every ten people will be involved in an alcohol-related crash at some point in their lives.

- On average, drivers impaired by alcohol cause 40 percent of all driving-related fatalities.

- Alcohol-related crashes are estimated to cost the U.S. economy at least $40 billion each year.

Facts About Drug Use

- Cocaine is a highly addictive stimulant that directly affects the brain.

- Addicts will not necessarily be convinced to stop using drugs even if you point out to them how their addiction is affecting their lives and the lives of their loved ones, and even if you inform them of all the harmful potential effects of drug abuse.

- Approximately 90 percent of property crimes and muggings are drug-related.

- On average, a person who is addicted to drugs needs about $200 each day to support the habit.

- Substance abuse can begin in children as early as elementary school.

index

Tools for Success
Critical Skills for the Construction Industry